Handbook of Basic Microtechnique

McGRAW-HILL PUBLICATIONS IN THE BIOLOGICAL SCIENCES

Consulting Editors
David M. Bonner, Robert L. Sinsheimer,
Colin S. Pittindrigh, Theodore H. Bullock

Series in Cell Biology
Gray, Handbook of Basic Microtechnique

Series in Developmental Biology
Patten, Foundations of Embryology

Series in Organism Biology
Mazia and Tyler, General Physiology of Cell Specialization

Series in Population Biology
Ehrlich and Holm, The Process of Evolution

Series in Systematic Biology
Hyman, The Invertebrates: Protozoa through Ctenophora (vol. I)
    The Invertebrates: Platyhelminthes and Rhynchocoela (vol. II)
    The Invertebrates: Acanthocephala, Aschelminthes, and Entoprocta (vol. III)
    The Invertebrates: Echinodermata (vol. IV)
    The Invertebrates: Smaller Coelomate Groups (vol. V)
Mayr, Linsley, and Usinger, Methods and Principles of Systemic Zoology

This symbol of the cephalopod Nautilus *appears on all McGraw-Hill Publications in the Biological Sciences. It was chosen to represent the just proportion of living structures and to suggest the harmonious workings and balanced arrangement of the parts and elements of living things. The color of the binding represents the Biological Series in which this book is published.*

Handbook of Basic Microtechnique

THIRD EDITION

PETER GRAY, Ph.D., D.I.C.

Department of Biological Sciences
University of Pittsburgh

McGRAW-HILL BOOK COMPANY

New York
San Francisco
Toronto
London

Handbook of Basic Microtechnique

Copyright © 1958, 1964, by McGraw-Hill, Inc. All Rights Reserved.
Copyright, 1952, by McGraw-Hill, Inc. All Rights Reserved.
Printed in the United States of America.
This book, or parts thereof, may not be reproduced in any form without permission of the publishers.
Library of Congress Catalog Card Number 63-22430

24206

Preface

This third edition is intended, as were the first and second, to supplement Gray's "Microtomist's Formulary and Guide," a work too large for practical use in elementary classes. It seemed both to the author and to his publishers, when these two works were first projected, that the field of microscopical technique presented an unusual dilemma. One horn was the necessity of providing an authoritative and exhaustive account of the main techniques developed since the invention of the art; "The Microtomist's Formulary and Guide" covers every branch of biological microtechnique and contains some 4,500 literature references. The other horn was the necessity of providing college teachers with a practical text that would cover the main requirements of undergraduate students in the fields of bacteriology, botany, zoology, premedicine, and medical technology. This horn proved much the sharper, and this third edition of "Handbook of Basic Microtechnique" bears little relation to the first and has been considerably amplified over the second. The author is happy again to have the opportunity of expressing to the many hundreds of teachers who have written to him his gratitude for the improvements that they have suggested and his regret that it is impossible either to acknowledge all these suggestions by name or to incorporate all of them in the present work. Such techniques as the preparation of fluid wholemounts, the complex metal staining techniques required to demonstrate much of the structure of the central nervous system, and the innumerable variations of triple staining are examples of techniques which are found in the larger work but which space limitations have kept out of the present volume.

The major change in the present edition, requested by almost all correspondents, has been the inclusion of a fourth part offering, in very abbreviated form, suggestions for both the source and the technique to be used in demonstrating a wide variety of anatomical, histological, and cytological details of both animals and plants. With very few exceptions the techniques recommended are those to be found in Parts Two and Three of the present work, though an occasional reference to "The Microtomist's Formulary and Guide" has been found necessary. The author is particularly indebted to his colleagues in the Department of Biological Sciences of the University of Pittsburgh for having criticized, and suggested additions to, this list. In particular, Dr. Malcolm Jollie, in the field of vertebrate anatomy, and Drs. Carl Partanen and Joan Gottlieb, in the field of botany, have saved him from making many mistakes, though they can in no way be held responsible for those which remain.

Another major alteration has been the inclusion in Part One of a new section on the theory and practice of phase-contrast microscopy and an explanation of dark-field microscopy. The whole of Part One, indeed, has been updated, and many of the illustrations have been replaced to show more modern equipment than that which was available when the second edition was published. The section dealing with setting up a medical school microscope with a separate lamp has been retained since the author's medical school colleagues report that many students still arrive with such instruments. However, a brief new illustrated section on the use of the microscope which has a built-in illuminating system has been added.

In the interests of space limitations, suggestions for supplementary reading material have been limited. The author's larger work not only is copiously furnished with references but contains a list of the 300 journals and 120 texts used in its compilation. In addition to this the author, in collaboration with his wife, has separately published a comprehensive classified and annotated bibliography ("Annotated Bibliography of Works in Latin Alphabet Languages on Biological Microtechnique," by Freda and Peter Gray, W. C. Brown Company, Dubuque, Iowa, 1956), which also contains a history of the field.

Finally, the author's special thanks are due to his secretary, Mrs. Janet Natowitz, for her unfailing and continuous help in the labor of producing this revised and expanded third edition.

Acknowledgment is made with thanks to the Fisher Scientific Company for Figs. 5-1 to 5-8, 12-5, 13-1, and 13-2, and to the American Optical Company for Figs. 12-9, 12-18, and 12-41.

Peter Gray

Contents

Preface   v

PART ONE: THE MICROSCOPE

Chapter 1.   Principles of Microscopy   3

    Lenses and Images
    Oculars
    Substage Condensers
    Light Sources
    How Images Are Seen

Chapter 2.   The Use of the Microscope   31

    The Freshman Microscope
    The Medical Microscope
    The Research Microscope

Chapter 3.   Photomicrography   55

    The Nature of the Photographic Process
    Processing Photographic Materials
    Photomicrographic Cameras
    Taking a Photomicrograph

PART TWO: THE PREPARATION OF MICROSCOPE SLIDES

Chapter 4. Types of Microscope Slides 77

Chapter 5. Materials and Equipment 80

Chapter 6. Fixation and Fixatives 85

Purpose of Fixation
Fixative Mixtures
Operations Accessory to Fixation

Chapter 7. Stains and Staining 97

Principles of Staining
Hematoxylin
Carmine
Orcein
Synthetic Nuclear Stains
Plasma or Contrast Stains
Stains for Special Purposes

Chapter 8. Dehydrating and Clearing 118

Chapter 9. Mounts and Mountants 124

Gum Media
Resinous Media

Chapter 10. Making Wholemounts 128

Temporary Wholemounts
Mounting in Gum Media
Mounting in Resinous Media

Chapter 11. Making Smears and Squashes   140

Smears
Squashes

Chapter 12. Making Sections   145

Nature of the Process
Free Sections
Paraffin Sections
Frozen Sections

Chapter 13. Cleaning, Labeling, and Storing Slides   191

PART THREE: SPECIFIC EXAMPLES
OF SLIDE MAKING

Example 1. Preparation of a Wholemount of a Mite by the Method of Berlese   197

Example 2. Preparation of a Wholemount of Pectinatella Stained in Grenacher's Alcoholic Borax Carmine   200

Example 3. Preparation of a Wholemount of 33-hour Chick Embryo, Using the Alum Hematoxylin Stain of Delafield   206

Example 4. Preparation of a Wholemount of a Liver Fluke, Using the Carmalum Stain of Mayer   212

Example 5. Smear Preparation of Monocystis from the Seminal Vesicle of an Earthworm   216

Example 6. Smear Preparation of Human Blood Stained by the Method of Wright 219

Example 7. Staining a Bacterial Film with Crystal Violet by the Technique of Lillie 222

Example 8. Demonstration of Gram-positive Bacteria in Smear Preparation by the Method of Gram 225

Example 9. Demonstration of Tubercle Bacilli in Sputum by the Technique of Neelsen 227

Example 10. Preparation of a Squash of the Salivary Glands of Drosophila Stained in LaCour's Acetic Orcein 230

Example 11. Preparation of a Transverse Section of a Root, Using the Acid Fuchsin–Iodine Green Technique of Chamberlain 233

Example 12. Preparation of a Transverse Section of the Small Intestine of the Frog Stained with Celestine Blue B–Eosin 237

Example 13. Preparation of a Transverse Section of a Stem of Aristolochia Stained by the Method of Johansen 245

Example 14. Demonstration of Spermatogenesis in the Rat Testis, Using the Iron Hematoxylin Stain of Heidenhain 250

Example 15. Preparation of a Transverse Section of the Tongue of a Rat, Using Celestine Blue B Followed by Picro Acid Fuchsin 255

Example 16. Preparation of a Transverse Section of Amphioxus, Using the Acid Fuchsin—Aniline Blue—Orange G Stain of Mallory   258

Example 17. Demonstration of Diplococci in the Liver of the Rabbit, Using the Phloxine—Methylene Blue—Azur II Stain of Mallory, 1938   261

Example 18. Preparation of a Series of Demonstration Slides, Each Having Six Typical Transverse Sections of a 72-hour Chick Embryo, Using the Acid Alum Hematoxylin Stain of Ehrlich   264

PART FOUR: RECOMMENDED TECHNIQUES

Animal Organs   271

Skin and Associated Structures
Nervous System and Associated Structures
Digestive and Respiratory Structures
Skeletal Structures
Glands
Circulatory Structures
Renal Reproductive Structures

Animal Tissues   277

Epithelial Tissues
Connective Tissues
Nervous and Sensory Tissues

Animal Cytology   279

Nuclei
Cytoplasmic Organelles

Plant Organs   280

> *Stem*
> *Root*
> *Leaf*
> *Flower*

Plant Tissues   282

Plant Cytology   284

Suggested Additional Sources of Information   286
Index   289

Handbook of Basic Microtechnique

PART ONE | The Microscope

CHAPTER 1 | Principles of Microscopy

The word microscope means "little seer" or "seer of little things." The first simple microscopes were glass globes filled with water, and Pliny has left a record of their use in the first century. No one, however, can examine the carved gemstones of antiquity without realizing that microscopes were in use at least five hundred years earlier. There is no theoretical limit to the magnifying power of such simple lenses, and Leeuwenhoek was able to make them so well that he discovered bacteria. He never saw what we now call a microscope.

LENSES AND IMAGES

There are many practical disadvantages to simple lenses of high magnifying power. The distortions they produce, which will be discussed later, are not the chief of these. The greatest objection is that focal length decreases as magnifying power increases, so that Leeuwenhoek had almost to push his eye into his lenses to see anything. This difficulty was overcome, about the year 1600, by using the newly invented telescope to examine from a reasonable distance the image made by a simple lens. This is exactly what we do today when we use what is now called a microscope, shown diagrammatically in Fig. 1-1.

The objective at the front end of the barrel is a powerful "magnifying glass" that produces an enlarged image of the object O at I_1. The extent to which this image is magnified is called the *primary magnification*, and nowadays this figure is usually engraved on the barrel of the objective. The lowest power on laboratory microscopes is usually $\times 3.5$ or $\times 4$. The highest is either $\times 45$ or, if there is an oil immersion, $\times 90$.

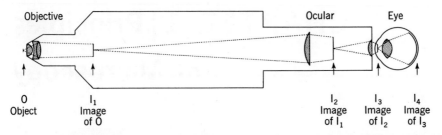

Fig. 1-1 | **Diagram showing production of image by a compound microscope.** A magnified image at I_1, thrown by the objective, is examined through a telescopic device known as the *ocular*, from which an image is formed on the retina by the lens of the eye.

This magnified image I_1 is examined through a telescope, known as the *ocular* or *eyepiece*, at the top end of the barrel. The first lens of the ocular produces a magnified image of I_1 at I_2, and the second lens of the ocular produces a small image—usually about a millimeter in diameter—of I_2 at I_3. This is a real image that may be demonstrated by holding a piece of translucent paper just above the eyepiece and lowering it until the small disklike image is sharply defined. This image is called *Ramsden's disk*.

Ramsden's disk is rather small, and it requires the human eye to transform it into the illusion of a magnified image of the object. This is done by advancing the eye until Ramsden's disk (I_3) is just inside the cornea. Under these circumstances the lens of the eye casts an image of Ramsden's disk over the whole surface of the retina at I_4. The image, in fact, "fills the eye." The extent of the apparent magnified image can be obtained by extending the dotted lines running from the lens to the retina as far as the plane of the object O. The production of a real magnified image on a photographic plate will be discussed later.

The distance between the outer surface of the top lens of the ocular and Ramsden's disk (I_3) is known as the *eye relief* of the ocular. In most oculars it is so stupidly short that wearers of spectacles have to remove them in order to get Ramsden's disk into the cornea and thus fill their eye with the image. A few manufacturers offer a limited range of oculars with long eye relief, and it is ardently to be hoped that these will in time become universal.

Aberrations of Lenses. The distortions of shape and color in the images made by simple lenses tormented the early makers of microscopes. These distortions are due to two simple facts illustrated in Fig. 1-2, which shows a section of a glass prism with a beam of light going

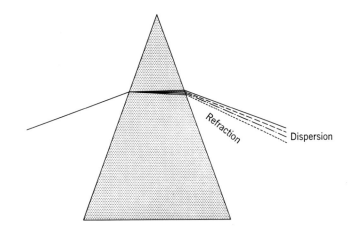

Fig. 1-2 | Dispersion and refraction by a glass prism. Refraction is a measure of the amount a beam of light is bent. Dispersion is a measure of the amount by which rays of different colors are bent.

through it. The beam, entering from the left, is of so-called "white" light. That is, it is a pencil of rays of mixed wavelengths which, when combined on the retina, cause the sensation that we have learned to call white. When a ray of light passes from a medium of one optical density to a medium of another—as from air to glass, or glass to water, or water to air—it is bent, or refracted. The extent to which it is bent depends first on the angle at which it enters the new medium and second on the difference in optical density between the two media. This difference is usually expressed as an index of refraction, which is the relative optical density of the medium in relation to air. Glass, for example, has about one and one-half times the optical density of air, so that its index of refraction, i in most books, is about 1.5. The value of i for water is about 1.3 and for a diamond about 2.4.

All these figures have been given as "about" so much because an accurate figure can be given only for a single wavelength, that is, a single color, of light. Each wavelength is bent a different amount, so that a pencil of white light passing through a prism, or lens, is dispersed as well as bent. This is also shown in Fig. 1-2. The difference between the index of refraction for red light, which is bent least, and violet light, which is bent most, is known as the *dispersion*. In glass and in most naturally occurring transparent substances, a high index of refraction and a high dispersion go together. For example, a diamond has a value for i of 2.41 for red light and 2.47 for violet light, giving a dispersion of 0.06. These facts, as will be apparent in a moment, make life very difficult for lens designers.

Now let us examine the formation of an image by a lens. Figure 1-3 shows a simple lens forming an image of an object. This object is an arrow

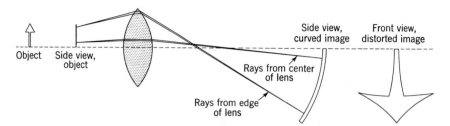

Fig. 1-3 | Diagram showing the causes of spherical aberration. Rays passing through the edge of the lens are bent more than those passing through the center. The rays passing through the edge of the lens therefore come to a focus nearer the lens, with the result that the image is curved. As those portions of the image that are nearer are larger, it follows that the image is distorted.

seen in front view at the far left of the picture, and in side view just to the left of the lens. Rays diverging from the lower part of the object to the lower part of the lens strike the lens at a relatively small angle and are therefore bent relatively little. It follows that they go a relatively long way before converging again to form the bottom part of the image. Rays diverging from the top part of the object to the top part of the lens strike the lens at a relatively large angle and are therefore bent more than the rays striking at a low angle. It follows that they go a relatively shorter distance before converging to form the upper part of the image. It is obvious that this effect is proportional all over the lens, so that the image seen in side view is curved. Actually, since the surface of the lens is part of a sphere, the image is also part of a sphere, and this effect is known as *spherical aberration*.

Now, the size of an image is dependent on the relative distance of the image from the lens. Hence spherical aberration produces not only an image that is curved but also an image that is distorted in shape when it is cast on a flat surface. This distortion is shown, exaggerated for the sake of clarity, in the front view of the image seen at the far right of Fig. 1-3. Spherical aberration is the result of refraction. Dispersion produces chromatic aberration.

Figure 1-4 shows exactly the same setup as Fig. 1-3 but rearranged to show the cause of chromatic aberration. Pencils of light diverging from the object are dispersed as they go through the lens. The red components are bent least and therefore travel a relatively long distance before converging to form a red image. The blue components are bent most and therefore travel a relatively short distance before converging to form a blue image. When these two images are seen in front view, the central portion, in which all the colors are superimposed, still appears white.

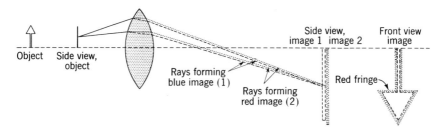

Fig. 1-4 | Diagram showing the causes of chromatic aberration. Light is dispersed (see Fig. 1-2) as well as refracted by the lens. The blue components are bent more than the red components, so the blue image is nearer the lens than the red image. There is, therefore, a red fringe around the image seen in front view.

The red image is, however, larger than the blue image and sticks out around it. The edges of this image therefore show up as a color fringe around the outside of the white image.

Correction of Aberrations. Spherical aberration is theoretically easy to correct since it is caused by the difference in distance between the center and the edge of a lens. The easiest method (Fig. 1-6) is to place a diaphragm in front of the lens so that only the center is used. The same effect (Fig. 1-7) is obtained from a cylindrical piece of glass with a lens face ground on each end. This, which is to all intents and purposes the center cut from a larger lens, is often sold as a hand magnifying glass. The least commonly used method (Fig. 1-5) is the so-called "meniscus lens," in which one face partially compensates for the other.

Figs. 1-5, 1-6, 1-7, and 1-8 | Methods of correcting aberrations. The meniscus (Fig. 1-5) diminishes the difference in thickness between the center and the edge. An iris diaphragm in front of the lens (Fig. 1-6) or a Coddington lens (Fig. 1-7) produces the same effect at the expense of the aperture. The action of the doublet (Fig. 1-8) in diminishing chromatic aberration is discussed in the text.

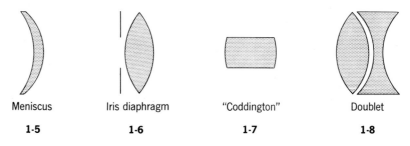

Combinations of all three of these ways of correcting shape distortions are used in microscope objectives.

The correction of chromatic aberration is much more difficult. The only solution so far discovered makes use of the fact that the relation between index of refraction and dispersion differs in different kinds of glass.

Take, for example, the combination shown in Fig. 1-8. If the positive lens at the left is made of a glass of high refractive index and low dispersion, it will bend the light a great deal and separate the colors very little. The negative lens on the right, if made of a glass with a lower index of refraction, will not bend the light out as much as the one on the left bends it in, so that an image will be formed. But if the negative lens on the right has a very high dispersion, it will pull the colors proportionately closer together, even though it does not bend the light so much. Hence all the colors will come together at the same point and produce a colorless, or achromatic, image. Moreover, this image will be relatively free of spherical aberration since the thick edge of the negative lens balances the thick center of the positive one.

Unfortunately, this is a theoretical dream. Glasses of very high refractive index and very low dispersion—or vice versa—do not exist. All lens design is a compromise, sometimes involving as many as six kinds of glass, each bending or separating or compressing light in varying amounts. Lens designers early learned to incorporate fluorite as a substitute for one of the glasses, and during World War II many synthetic nonsilica glasses were developed. But the perfect lens is still in the future. Designers effect the best possible compromise between reality and theory in the light of the specific requirements of a microscope objective. These requirements must next be examined.

Resolution. The most important part of a microscope is the objective. Every other part of the instrument is designed to help the objective produce the best possible image. The best image is not the *largest*—it is the *clearest*. There is no purpose in looking at an object through the microscope unless we arrive at a better understanding of its structure. Mere size is no aid to understanding. A simple black dot the size of a pinhead is just as understandable as a simple black dot an inch in diameter. What we want to know from the microscope is whether the pinhead-sized dot is a simple dot or whether its smallness conceals a pattern. The ability of the microscope to reveal this pattern is known as *resolution*, and resolution is therefore the most eagerly sought characteristic of a lens. It is obvious that a lens with chromatic and spherical distortions will not resolve satisfactorily, but there is more to resolution than the correction of aberrations.

The two major factors in resolution are the wavelength of light used and the angular aperture of the lens system involved. The effect of wavelength is the easiest to explain and will be left until later; angular aperture is a difficult concept and will be tackled first. Light which crosses from a substage condenser through an object to an objective (Fig. 1-10) is in the form of a cone or, as it is commonly called, a *pencil* of rays. When this cone has a relatively narrow angle, like the tip of an actual pencil, it does not spread out much after it has passed through the object. Since it does not spread out, it does not separate by much the images of closely crowded objects through which it has passed. If, on the contrary, the cone has a very great angle and spreads out rapidly, closely crowded objects will appear widely separated and will be resolved. A crude analogy would be to suppose that the student had in his hands a closely woven fabric with a mesh so fine as to be scarcely perceptible to the eye. If he spread it slightly, the individual fibers would still be imperceptible; but if he spread it greatly, it would be resolved into its component parts.

The effectiveness of angular aperture is, however, limited in practice by the wavelength of the light used. Again it is necessary to ignore theoretical arguments as to the structure of light rays in favor of a practical analogy. Suppose that light is propagated as a series of waves of varying wavelengths. Now mentally transpose these waves to the surface of the ocean. A liner will leave a perceptible wake—or, in optical terms, shadow—as it passes through even the largest waves. A child's toy boat will leave a wake of only the tiniest of ripples. Light rippling past an object on its way through a microscope to our eyes follows just the same rules. No object smaller than the waves of light can create a disturbance in the waves that is perceptible to the eye. It follows that the shorter the wavelength of light, the greater the possible resolution.

It must be reemphasized at this point that resolution, from the practical point of view, is a measure of crispness or clarity. People often fail to see how the number of lines or dots per inch that can be separated, or resolved, is a measure of the sharpness with which larger objects can be seen. Actually everything is seen against a background of something else. Sharpness and clarity are just measures of how well the object is separated from, or resolved against, the background.

Wavelength, resolution, and angular aperture have very simple relationships. In the first place it must be obvious that the angle of the cone of light that can enter any lens is dependent on the refractive index of the medium in which the lens is working. This dependence, or numerical aperature, is expressed by the relation

$$\text{N.A.} = i \sin \theta$$

where θ is one-half the angle of the entering cone of light and i is the refractive index of the medium surrounding the lens. Since i for air is 1, and since $\sin \theta$ cannot be greater than 1, it follows that no lens working in air can have a theoretical N.A. greater than 1.

The relation between N.A. and resolution is just as simple for

$$R = \frac{\text{N.A.}}{\lambda}$$

where λ is the wavelength of the light and R is the number of lines that can be separated. R or λ can be given in either inches or millimeters. Some books prefer to express resolution in lines per inch and wavelength in millimeters, in which case the relation becomes

$$R = \frac{2.54 \text{ N.A.}}{\lambda}$$

The figure for N.A. that is engraved on the barrel of an objective is therefore an indication of the maximum resolution of which the lens is capable. It is *not*, and this cannot be reiterated too often, a measure of the resolution that the lens will automatically produce. It is only a measure of what the objective can be made to yield in the hands of a skilled microscopist provided with all the necessary auxiliary equipment.

Figures 1-9 to 1-11 will make this clear. Figure 1-9 shows the setup with an ordinary elementary class microscope. The objective is focused from above through a coverslip on an object on a slide. Some distance under the slide is a concave mirror reflecting a cone of light into the objective. The maximum angle of this cone is dependent on the size and distance of the mirror and on nothing else. In the example shown, drawn to scale for a standard microscope, this angle is 64°. Since the lens is working in air ($i = 1$), it follows that the maximum numerical aperture of the system is

$$\text{N.A.} = i \sin \theta = 1 \times \sin 32° = 0.6$$

It does not matter in the slightest what N.A. figure is engraved on the barrel of the lens. The maximum possible N.A. of the system is 0.6, and

Principles of Microscopy 11

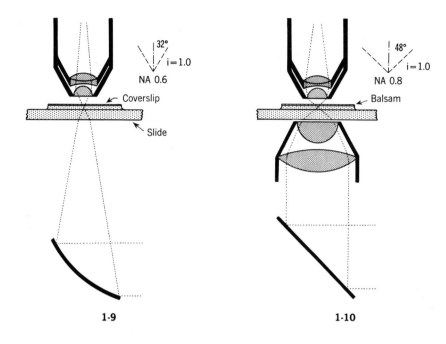

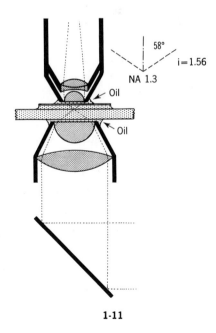

Figs. 1-9, 1-10, and 1-11 | Diagrams showing relation between angular aperture and resolution. Fig. 1-9 shows the concave mirror of a freshman-type microscope (Fig. 2-1) casting a narrow cone of light into the objective. This cone has a total angle of 64° so that the numerical aperture is 0.6. Fig. 1-10 shows a substage condenser used to increase the angular cone to 96°. Since the system is working in air ($i = 1$), the numerical aperture is 0.8. Fig. 1-11 shows both the substage condenser and the objective oiled to the slide. The angle of the cone is 116°, and i, for the oil, $= 1.56$. The N.A. is thus 1.3. The relation between N.A. and resolution is explained in the text.

this cannot be increased unless the angle of the cone is increased. This is the function of the substage condenser.

A substage condenser, shown in position in Fig. 1-10, is, in effect, a large objective with a very high numerical aperture. Its function is to project a wide-angle cone of light through the slide into the objective. As this condenser is working in air, it obviously cannot have an N.A. greater than 1. In the example shown, the full aperture of the condenser is not being used, so that the cone of light entering the objective has an angle of 96°. It follows that the working aperture is

$$\text{N.A.} = i \sin \theta = 1 \times \sin 48° = 0.8$$

Even at the risk of appearing tedious, it must again be emphasized that the objective may have N.A. 0.95 engraved on the barrel, and the condenser may have N.A. 1.4 engraved on the top lens housing. This has nothing to do with the case. The working N.A. is dependent on the actual cone of light being projected, which is in its turn determined by the aperture in the iris under the substage condenser.

The addition of the wide-angle substage condenser can raise the working N.A. close to, but never beyond, 1. The only way of achieving an N.A. greater than 1, since $\sin \theta$ cannot rise above this figure, is to increase i. This is the reason for the use of immersion oil and of immersion lenses made to work in it. This situation is shown in Fig. 1-11. Immersion oil, with a refractive index of 1.56, has been substituted for air both above *and below* the slide. This not only increases i but also permits the lenses involved to operate at a greater angular aperture. In this case

$$\text{N.A.} = i \sin \theta = 1.56 \times \sin 58° = 1.3$$

It should particularly be remarked that if oil is used only between the objective and the slide, the system cannot operate at an N.A. greater than 1. This may not in practice be a bad thing, but this will be discussed later. In the meantime, it is necessary to turn to other requirements of objectives.

Magnification. Magnification is, of course, a requirement of an objective even though it is entirely secondary to resolution. The useful limit of magnification, in fact, is that which increases the size of the smallest object that can be resolved to the smallest object that can be seen. Assuming, roughly speaking, that the eye can just see a speck 0.1 mm in diameter, the useful limits of magnification in relation to N.A. of common types of objectives are given in Table 1.

TABLE 1

"Equivalent focus" and primary magnification of typical objective	Numerical aperture of objective	Smallest object which can be resolved (angstrom units)	Magnification required to bring object to apparent size of 0.1 mm
2 mm ×90	1.4	3,600 Å	×2800
	1.3	3,800 Å	×2600
	1.2	4,200 Å	×2400
4 mm ×45	0.95	5,300 Å	×1900
	0.85	5,900 Å	×1700
	0.65	7,700 Å	×1300
8 mm ×20	0.60	8,300 Å	×1200
	0.50	10,000 Å	×1000
16 mm ×10	0.30	16,500 Å	×600
	0.25	20,000 Å	×600
30 mm ×3.5	0.10	50,000 Å	×200
48 mm ×2	0.08	62,500 Å	×125

The total magnification is, of course, a multiplicand of the magnifications of objective and ocular. Even so, it will immediately be seen that the usually employed ×10 eyepiece does not result in magnifications anywhere near the useful limit of most lenses. These limits are, however, often exceeded in photography, where actual magnification can be increased at will by increasing the distance between eyepiece and film. Moreover, the residual aberrations of most lenses cause the image to degenerate long before the theoretical limit of magnification is reached. In general practice, the old rule that "the useful limit of magnification is one thousand times the N.A." is a useful one to follow. For example, a ×45 objective working at an N.A. of 0.85 will be stretched beyond its useful limit with a ×20 eyepiece but will probably stand a ×15 eyepiece.

Two methods of designating the primary magnifications of objectives are still in use. The obviously intelligent one, followed by most contemporary manufacturers, is to engrave the actual figure on the lens barrel. The older method, still all too frequently seen, and given for comparison in the table of magnifications, is to engrave the equivalent focus. When this equivalent focus is given in inches—e.g., ½ in., ⅔ in., ⅙ in.—it has some actual meaning. The equivalence, in this case, is to the magnification of a simple lens at a distance of 10 in. A simple lens of 1 in. focal length will cast an image magnified ten times at a distance of 10 in. Similarly a ½-in. lens will give a ×20 image at the same distance, or a

⅔-in. lens a ×15 image. It is therefore relatively easy, when a lens has E.F. ⅙ in. engraved on the barrel, mentally to convert to ×60. The engraving of E.F. in terms of millimeters is a stupidity that should never have originated, let alone be used by American manufacturers. To interpret, for example, E.F. 2 mm, it is first necessary to convert 2 mm to ¹⁄₁₂ in. before arriving at ×120. Actually most so-called "2-mm" lenses have a primary magnification of ×90, which reduces the whole system to imbecility.

Working Distance. The working distance of an objective is the actual distance between the front surface of the lens and the coverslip. This has no relation to the equivalent focus since it is just as dependent on N.A. as on magnification. Reference back to Figs. 1-9 and 1-10 will make this clear. It is impossible to increase the theoretical N.A. of a lens without widening the angle of the entering cone of light. It is impossible to increase the angle of the cone without pushing the lens closer to the coverslip. The working distance of a lens of large N.A. is therefore very short no matter what the actual N.A. at which the system is operating. This is not of much importance with low-power lenses but is critical with high-power lenses such as the high-dry ×45 (the old 4 mm). Many student microscopes of the type shown in Figs. 1-9 and 2-1 are furnished with high-dry lenses of N.A. 0.85. These lenses cannot possibly be used at an N.A. greater than 0.6, so the higher N.A. has no purpose except to ensure a large number of broken coverslips at the hands of a beginning student.

Working distance also decreases as actual image magnification increases. It is almost impossible, for example, to use a ×45 N.A. 0.95 objective on a microprojector since the huge magnification of the projected image requires a working distance less than the thickness of an average coverslip.

Working distance, in short, is a very important practical factor in the selection of objectives. Very expensive lenses of high magnification and large N.A. may occasionally be necessary in research, but they have no place in teaching or routine laboratory work.

Correction of Objectives. The methods by which corrections of lenses are made have been discussed earlier. The compromises mentioned there have resulted in two series of lenses known as *achromats* and *apochromats*. The words mean, respectively, "without color" and "with separated colors."

A theoretically perfect achromat, which does not exist, would bring all wavelengths of light to a common focus so that, in terms of Fig. 1-4, all the images would be of the same size. This is not practically possible, so an endeavor is made to bring together those colors—red and green—to which the eye is particularly sensitive. The residual blue color fringe of the image is not readily detected by the eye and vanishes completely against a blue background. For this reason, manufacturers commonly provide a disk of blue glass in front of cheap microscope lamps. Green glass has exactly the same effect and is far more restful to the eye.

Achromatic lenses are relatively cheap and are adequate for most purposes. They are usually made with relatively low numerical apertures and long working distances.

Apochromatic objectives are made on an entirely different principle and must be used with special eyepieces. The colors are intentionally widely separated—or overcorrected—and then brought back to a nearly common focus by a compensating eyepiece. An apochromatic objective used with an ordinary eyepiece gives a much worse image than the much cheaper achromatic. Apochromatic objectives are available only in higher magnifications, usually from $\times 20$ up, with very high numerical apertures and correspondingly short working distances. Their use is justified only in photography, particularly color photography, and in the most critical research.

OCULARS

There are, as indicated in the last paragraph, two main series of oculars: the regular, or Huygenian, ocular for use with achromatic objectives and the compensating ocular for use with apochromatic objectives. Each of these series is available in a number of powers, or magnifications, of which the $\times 10$ is standard with most manufacturers. Oculars of lower power than this are of little practical value, except on the rare occasion when an increased field of view is required. Oculars of higher power, particularly the $\times 12.5$ and $\times 15$, are often useful to reduce eyestrain when making continued examinations of small objects. It must be remembered that these oculars add nothing to the resolution of the system but only increase the apparent size of objects already resolved by the objective. The salivary chromosomes shown in Fig. E10-1 are a good example. A good $\times 20$ (8-mm) objective clearly resolves even the finest bands on the chromosomes, but it is irksome to count and compare the bands at a magnification of $\times 200$. The substitution of a $\times 15$ eyepiece to raise the total magnification to $\times 300$ adds nothing to the knowledge, but a great deal to the comfort, of the observer.

Too much attention should not be paid to flatness of field for visual observation. The eye observes only a small portion of the field of view at one time, and a touch on the fine adjustment will bring the periphery into focus. Flatness of field is, however, vital in photomicrography, and the special eyepieces designed for this use are discussed in the next chapter.

SUBSTAGE CONDENSERS

The substage condenser is just as vital to resolution as the objective. Broadly speaking, two types are in use today—the Abbe and the achromat. This is rather confusing since the first type is commonly employed with achromatic objectives and the second with both apochromatic and achromatic objectives. An apochromatic condenser is not, of course, a feasible proposition since there is no possibility of inserting into the system a supplementary compensating ocular to draw the widely split rays together again.

The critical criteria of substage condensers are maximum N.A., working distance, and focal length. The term maximum N.A. is used advisedly, because every substage condenser is furnished with an iris diaphragm, *the only function of which is to control the N.A.* It is a widespread delusion, ruinous to good microscopy, that this iris may be

Figs. 1-12, 1-13, and 1-14 | **Diagrams showing the effect of the substage iris on numerical aperture and on the size of the image of the light source.** Fig. 1-12 shows an Abbe condenser with the iris wide open casting a small image of the light source with a light cone of 83° and a consequent N.A. of 0.7. Fig. 1-13 shows the same condenser with the iris partially closed casting the same size image of the light source but with an angular aperture of 28°, so that the N.A. is now 0.2. Fig. 1-14 shows an achromatic condenser with the top lens removed casting a large image of the light source, so that a low-power objective can be used.

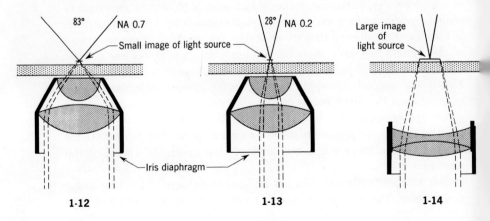

used to control the intensity of illumination. Reference to Figs. 1-12 and 1-13 should make this point clear. Figure 1-12 shows an Abbe condenser, with the iris almost fully open, throwing a cone of light with an 83° angle. Any lens placed above this will therefore be functioning, if it is constructed to do so, at an aperture of N.A. 0.7. Figure 1-13 shows the same condenser with the iris diaphragm partially closed. The emergent cone of light now has an angle of 28°, and any objective placed above it cannot operate at more than N.A. 0.2, no matter what figure is engraved on its barrel. It is true that decreasing the N.A. has the effect of diminishing the light intensity, but at the cost both of losing fine detail and, in extreme cases, of introducing detail that is not there. This is well seen in Figs. 1-15 to 1-17. Figure 1-16 is a photograph of the jagged torn edge of an extremely thin layer of silver foil. This is an excellent test object because there is no question of what it ought to look like. It ought to appear as a clean silhouette with crisp edges. This photograph was taken with an apochromatic ×45 (4-mm) objective operating at an aperture of 0.9. The photograph was then enlarged ten times to render obvious in reproduction details on the absolute limit of resolution of both the lens and the observer's eye. The minute speck of detached silver (1) thus appeared on visual observation as a just visible particle. Notice also that the leg-shaped projection to the right of this particle has a tiny bump in the sole of the foot. At (2) there is a narrow, but relatively deep, tear in the metal.

Now turn to the lower figure (Fig. 1-17). The only difference in the setup for taking this photograph was that the iris of the substage condenser was shut down until the lens was operating at N.A. 0.2. Notice first that the fine particle (1) and the bump in the sole of the footlike projection have vanished. They are, like the deep tear at (2), beyond the limit of resolution of a lens of N.A. 0.2, and the fact that the objective in question has N.A. 0.95 engraved on the barrel has nothing to do with its performance. Not only, however, has closing the substage iris caused fine detail to vanish or become distorted, but it has introduced many details that are not there. Notice particularly the appearance of nonexistent "vacuoles" in the solid metal. Those who believe, on the basis of the entirely false analogy of a camera lens, that "closing the iris brings out the detail" are likely to come up with some staggering inaccuracies of observation.

The top photograph (Fig. 1-15) shows the result of opening the substage iris too wide. The substage condenser used for these photographs had a maximum aperture, since it was joined to the slide by a drop of oil (Fig. 1-11), of N.A. 1.4. The cone of light entering the objective was therefore wider than the objective could handle. In consequence of this there was much internal reflection from lens barrel and microscope tube.

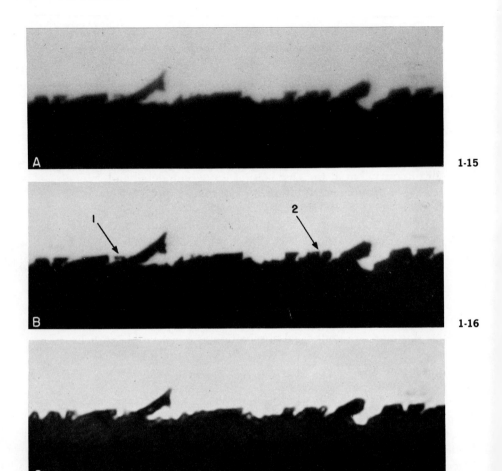

Figs. 1-15, 1-16, and 1-17 | The effect of numerical aperture, as controlled by the substage iris, on resolution and glare. Fig. 1-16 (B) is taken with a ×45 apochromatic objective working at N.A. 0.93 at a magnification of 1,500. The object shown is the torn edge of a thin sheet of silver foil, and the speck at 1 or slit at 2 lies at the absolute limit of resolution possible at this N.A. Closing the substage iris to produce an N.A. of 0.2 [Fig. 1-17 (C)] not only causes the speck and slit to vanish but inserts a lot of nonexistent vacuoles into the picture. Opening the iris to give an emergent cone of N.A. 1.3 into a lens working at N.A. 0.93 [Fig. 1-15 (A)] produces glare that masks the image.

These internal reflections produce glare, which makes it impossible to see, and very difficult to photograph, fine detail. In sum, the critical adjustment of the substage iris is one of the most important factors in the use of the microscope. The method of making these adjustments in practice is described in the next chapter. The present discussion has been for the sole purpose of emphasizing that the substage iris controls resolution, *not* light intensity. Before passing to sources of illumination, it is necessary to draw attention to another criterion of the substage condenser—working distance and focal length.

Figures 1-9 to 1-14 show clearly that the substage condenser focuses the light on the object. Actually, as will be explained in the next chapter, it focuses an image either of the light source or of an iris in front of the light source. The size of this image is naturally dependent both on the focal length of the condenser and on the distance of the light source from the condenser. A very high N.A. obviously necessitates a short working distance, which results in a short focal length and a very small image. This image is worthless if it does not fill the field of view of the objective, so there are objections as well as advantages to achromatic N.A. 1.4 condensers.

The ordinary Abbe condenser, with a maximum working N.A. of about 0.9, is perfectly adequate for most work. The relatively long focal length gives a working distance sufficient to permit focusing through even the thickest slide and produces an image of an adequate lamp sufficiently large to fill the field of a $\times 10$ (16-mm) objective. The resolution of an oil-immersion objective of N.A. 1.2, working at N.A. 0.9 with an Abbe condenser, is good enough for all student uses and for routine laboratory work.

The N.A. 1.4 achromatic condenser is required only when the highest possible color correction is sought with apochromatic objectives or when oil-immersion lenses are to be used at numerical apertures greater than 1. In this case, of course, the condenser, as well as the objective, will have to be oiled to the slide. The image of most light sources produced by these high-power condensers is too small to fill the field of even medium-power objectives, and there are only two ways to get around this. Either the light must be brought closer or the focal length of the condenser increased. The former is what in effect happens either with a built-in light source or with the American Optical illuminator described in the next section. The focal length of the condenser can be increased in two ways. The simplest, but least satisfactory, is to insert a low-power lens into the system immediately under the condenser. Many microscopes are fitted with either a swinging or sliding lens for this purpose. Unfortunately, these lenses not only expand the beam but also throw it out of focus, so that the corrections and resolution of the low-power objectives are seriously affected. The proper function of these swinging or sliding lenses is to permit a field to be searched with low power to locate an object for examination under high power.

The proper setup for critical microscopy with low- or medium-power lenses is shown in Fig. 1-14. Almost all achromatic condensers have the top lens mounted in a knurled ring. This top lens can thus be easily unscrewed and removed without seriously affecting the corrections of the remaining lenses of the system. The lower-power condenser so produced usually has an N.A. of about 0.5 and produces an image of the light source which will fill the field of a $\times 5$ (32-mm) objective. A few achromatic condensers are made so that the two top lenses of the system can be removed, but this is necessary only for low-power photomicrography, for which purpose it is usually better to buy a specially built condenser.

LIGHT SOURCES

The light source, whether built in or not, is actually an integral part of the microscope system. It is impossible to get the best service from a first-class microscope with a second-class illuminator. A first-class illuminator improves a second-class microscope almost beyond belief. It is, therefore, pathetic that people who will cheerfully spend the price of an expensive automobile on a microscope will object to buying a lamp costing less than a bicycle.

A student microscope, with an illuminating system of the type shown in Figs. 1-9 and 2-1, can operate satisfactorily with the light from a gooseneck lamp fitted with a frosted or opal bulb. The large area of the illuminating source and the low power of the condensing mirror combine to produce adequate coverage for the field of a $\times 3.5$ (32-mm) objective. The standard $\times 40$ (4-mm) N.A. 0.65 fitted to these instruments will function with this setup at an N.A. of about 0.30, which is just adequate for the crudest type of biological observation. Numerous "illuminators," consisting of a sheet of ground glass in front of a clear bulb, are on the market, but they offer no advantage over an opal bulb. Some of these devices are mounted on fancy adjustable stands and incorporate, for reasons apparent only to the manufacturer, an iris in front of the ground glass.

None of these devices is of very much use with a microscope incorporating a substage condenser since they do not permit the mirror (Figs. 1-9 to 1-11) to reflect a beam of light large enough to fill the back lens of the condenser. This has exactly the same effect as closing the iris (Fig. 1-13) and thus removes the whole reason for using a substage condenser.

The minimum requirements for a separate microscope illuminator for use with a microscope fitted with a substage condenser are:

1. A high-intensity lamp with a compact filament. This may be either a 110-volt projection bulb or a 6-volt bulb worked through a transformer.
2. A lens at least 3 in. in diameter, mounted in a focusing device, which will produce a sharp image of the filament at a distance of from 6 to 12 in.

3. An iris diaphragm in front of the lens.
4. A filter holder in front of the iris.
5. A housing that can be tilted.

Most manufacturers can provide a lamp to these specifications, and that shown in Fig. 2-8 is typical of the best of them. It must be emphasized that such an illuminator, the provision of which adds 25 per cent to the cost of a medical school microscope, and 10 per cent to the cost of a research microscope, is absolutely essential. There is no point in buying a ×90 N.A. 1.2 oil-immersion objective and an N.A. 0.9 or higher substage condenser if the system cannot be used at better than N.A. 0.6. Just as large and as well-resolved an image can be obtained with a ×45 N.A. 0.6 objective and a ×20 eyepiece at less than half the price.

The method of using the illuminator is described in the next chapter, but the reasons for its parts can be given here. An ideal image is produced when all those rays that leave the light source at one time also reach the eye at one time. It follows that an ideal light source would be a point. A compact filament is a compromise between theory and practicality. The lens is required to gather the rays leaving the filament and condense them into a beam of useful size and direction. This lens, as will be seen later, is in practice the light source of which the substage condenser forms an image on the slide. The iris in front of it is therefore required to limit the size of this image. This brings up the very important point of how to control the intensity of illumination, since neither the substage iris nor the lamp iris can be used for this purpose. There are two methods. The first is to vary the voltage applied to the lamp. This is simple but has the disadvantage that the color of the light becomes redder as the intensity is diminished. Much the best method is to place neutral density filters in the filter holder in front of the iris.

The high cost of providing the necessary illuminator and the difficulty of persuading the customer of the necessity for it are causing more and more manufacturers to turn to built-in illuminators. Those attached to the underside of the substage condenser are not very satisfactory, both because of the heat generated and because the cramped situation makes it difficult to provide an adequate condensing system. Those instruments, however, that have the illuminating system built into the stand are, in general, excellent.

The most satisfactory microscope illuminant on the market at present is the American Optical (formerly Silge and Kuhne) Ortho-Illuminator (Fig. 2-22). This relies on the new idea of using a series of fluorescent glass disks as the source of illumination. Three such disks, mounted on a swiveled plate, provide three practical light intensities of unvarying color. Red, green, and daylight filters, also on a swiveled plate, may be rotated in front of any of them. Both fluorescent disk and filter may be rotated out of the way to permit direct centering of the lamp filament. Once a microscope has been set up on this device, it is necessary only to switch on the lamp to have three possible intensities of three colors perfectly adjusted at all times.

HOW IMAGES ARE SEEN

All that has so far been said deals with the formation of images. It is just as essential that the student understand how these images are seen.

The familiar idea that "light is propagated as a sine wave" requires a great deal of modification and explanation. Light does not, even though it is almost always shown that way in diagrams, consist of a continuous wave. It is better, though not very much less misleading, to regard light as a series of small snakelike particles moving at the rate of 186,000 miles per sec. A "beam" of light consists of millions of these particles which are so close together that it is simplest to draw them as continuous waves. Each wave is a sine curve and has therefore, when drawn as though it were flat (top of Fig. 1-18), two important dimensions. These dimensions are the amplitude, or depth of the curve, and the wavelength, or distance between the two crests. The eye interprets amplitude

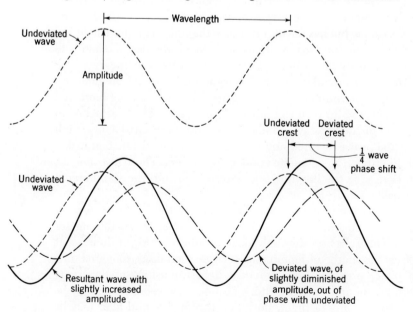

Fig. 1-18 | Diagram showing relation of light waves in normal illumination.

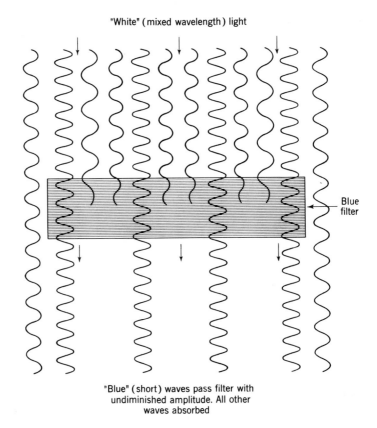

Fig. 1-19 | Diagram of light waves passing through a transparent blue filter.

as brightness and wavelength as color. Both these are purely subjective phenomena which are learned in early childhood. Neither, for example, can be explained to an individual who has been blind from birth. This early childhood training causes an individual on whose retina long wavelength light is falling to say that he is perceiving a red color, while he refers to the sensation of receiving short wavelength light as seeing a blue color. Green and yellow are intermediate wavelength lights. Most extraordinary of all, an individual is taught that a mixture of a great many wavelengths has no color at all—that it is white. The standard mixture of wavelengths thought of as white is that of the sun's rays which are reflected from a cloud in the sky.

The majority of objects seen under the microscope are transparent and are seen only if they contrast with the background in either intensity or color. Figure 1-19 shows a mixture of wavelengths representing white

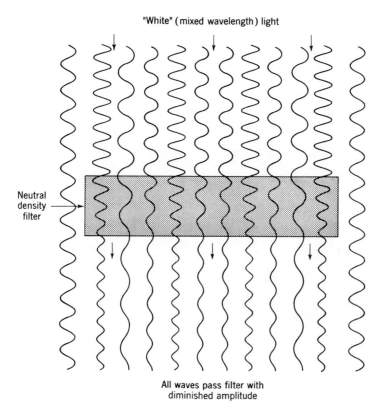

Fig. 1-20 | Diagram of light waves passing through a gray, or neutral density, filter.

light falling on a transparent blue substance. Short wavelengths representing blue light pass through without interruption, while all the other wavelengths are absorbed. A piece of transparent blue material in the field of the microscope therefore appears blue against a white background because the other components of the white have failed to get through it. Figure 1-20 shows a neutral-colored or gray object under the same circumstances. All the wavelengths which fall on it pass through but with a diminished amplitude, so that it does not appear so bright as the background and is therefore seen as a darker object.

It is very important at this point to distinguish clearly between what the microscopist refers to as the *deviated* and the *undeviated* rays which travel through the optical system of the microscope. All the previous diagrams in this book have shown only the deviated beam. Figure 1-21 shows light passing through a part of the lens system of a microscope in the field of which no object is in view. When these light rays are gathered

Principles of Microscopy 25

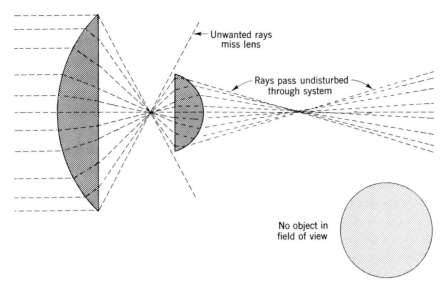

Fig. 1-21 | Diagram showing passage of light through a lens system without any object in the field and therefore without deviation of the beam.

Fig. 1-22 | Diagram showing passage of light through a lens system with an object in the field of view which deviates part of the beam. The deviated beam forms the image.

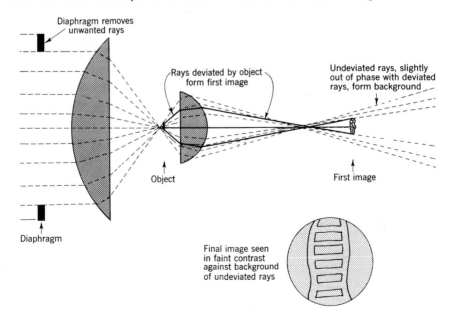

together by, and viewed through, the ocular, a simple uniformly illuminated circular field, as shown at the right of the figure, is seen. If, however, an object is placed in the field of view (Fig. 1-22), the light which touches this object is deviated, and the function of the front lens of the system is to bring this scattered light to a focus (compare Fig. 1-1) forming a primary image which is then examined by the ocular. The light which forms the image in the field of view is thus the deviated beam, while the light which passes around the object is the undeviated beam which forms the background against which the image is seen. An unstained object can be seen only when the amplitude of the light passing through it, *in relation to the background of undeviated rays*, differs from the undeviated rays in amplitude. There are two ways in which this can happen. One, the simple diminution of amplitude from the presence of a neutral color, has already been mentioned. The second, known as *phase shift*, is far more important.

As has already been pointed out, the waves representing the light are not stationary but are actually moving forward through the object. They do not move forward at the same speed through objects of different optical densities. If two waves of light are progressing exactly parallel to each other, so that the crests and the troughs of the waves remain exactly parallel as the light moves forward, the two waves are then said to be *in phase*. If, however, one of the waves is speeded up or slowed down, so that its crests and troughs no longer correspond with those of the other wave, it is said to be *out of phase*. The lower part of Fig. 1-18 shows what happens in this case. The dotted line shows the undeviated wave, which is identical with the one shown at the top of the figure. The dashed line represents the deviated wave (see Fig. 1-22), which has not only been diminished slightly in amplitude in its passage through the object but has also been thrown out of phase to the extent of about a quarter of a wavelength, so that it is no longer parallel to the undeviated wave. Remember that these two waves are superimposed on each other when observed through the eyepiece of the microscope. The wave which results from superimposing any two sine waves is arrived at by adding the positive (above the center line) and negative (below the center line) vertical components at any given point along the wave. The heavy black line shows the result of adding the vertical amplitudes of these two waves together and is the actual wave which produces the effect on the eye. It will be seen that it is slightly greater in amplitude than either of its components and is therefore seen as a slightly brighter light. The viewer seeing a particle in the field of the microscope (Fig. 1-22) which appears slightly brighter or *more refractive* is, in point of fact, seeing the effect produced in the lower half of Fig. 1-18.

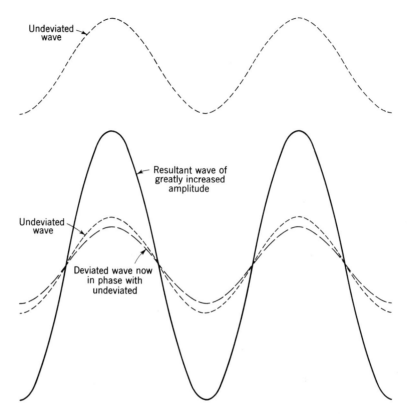

Fig. 1-23 | Diagram showing relation of light waves in phase-contrast illumination.

It must be obvious that this brightness will be enormously increased if a method can be devised to superimpose the deviated and undeviated waves on top of each other. This effect is known as *phase-contrast microscopy*, in which the undeviated beam is slowed up until it is again exactly parallel to the deviated beam. In adding the two waves together, all the negative and positive amplitudes are paired, so that the amplitude of the resultant wave is doubled.

Figures 1-23 and 1-24 show the principles by which phase-contrast images are seen. Turn first to Fig. 1-24. Underneath the substage condenser, the iris diaphragm is replaced by an annular stop. This is a disk which is opaque except for a clear ring. This clear ring therefore passes a hollow cylinder of light which becomes a hollow cone as it goes through the lens system. Those rays in this hollow cone which are not deviated by striking the object are brought to a focus on a phase plate;

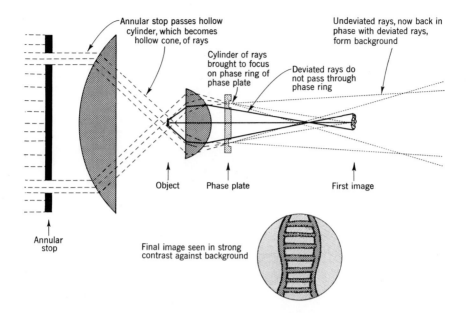

Fig. 1-24 | **Diagram showing the passage of light rays through a phase-contrast system.** The undeviated beam passes through the phase ring on the phase plate. The deviated beam passes through the plane portion of the phase plate.

that is, the undeviated beam produces an image of the annular stop on the phase plate. The phase plate is completely transparent but has on it a ring of material which shifts the phase of the light going through it by approximately a quarter wavelength. Those parts of the hollow cone of light which are deviated by striking the object are also themselves retarded by the object to the extent of about a quarter wavelength. These rays, however, do not go through the phase ring on the phase plate but continue without a further change in phase through the transparent parts of this plate. It follows that the deviated and undeviated waves are again brought back into phase, as shown in the lower half of Fig. 1-23. Since the crests and the troughs of both waves are now parallel to each other, the resultant wave, which is the sum of the two waves, will be doubled in amplitude. The apparent brightness of light is actually proportional to the square of the amplitude, so that the image seen in a phase-contrast microscope is four times as bright as that seen in a regular microscope. The effect here produced is known as *bright-phase* microscopy, but it is just as easy to have a *dark phase*. If the deviated beam is so shifted in phase that the crest of this beam corresponds to the trough of the undeviated beam, the resultant of the two will be halved in amplitude and the particle will therefore appear four times darker in the phase-contrast microscope than in the regular microscope.

The advantage of phase-contrast microscopy is that transparent objects continue to be seen as transparent objects even though the contrast between them and the background is increased. Another, and much older, method of increasing the contrast of the object against the background is known as *dark-field microscopy*. This is produced by the methods shown in Fig. 1-25. A solid central stop is placed underneath the substage condenser so that only the outside of the beam passes through. These outside rays—refer back to Fig. 1-22—do not enter the front lens of the objective unless they are deviated by the object. It follows that no part of the undeviated beam, which normally forms the background, is seen on looking through the eyepiece, so that the image appears in brilliant white against a black background. Unfortunately this method produces a great deal of glare, so that, to all intents and purposes, the object is seen as a brilliant silhouette rather than as a brilliant object in which detail can be made out. This method is still very useful for looking at such things as flagella on living bacteria or for

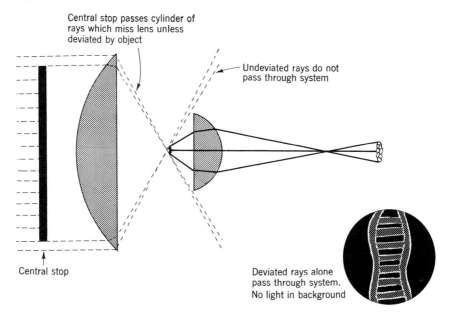

Fig. 1-25 | Diagram showing the passage of light rays through a dark-field system in which only the deviated beam passes through the objective.

determining the existence in sections of particles of the absolute limit of the resolution of the system. However, the first type of object is very nearly as well seen by phase-contrast microscopy, and the latter type is now rather unimportant since electron microscopy permits the study of such particles in detail.

SUGGESTED ADDITIONAL READING

Bennett, A. H., et al.: "Phase Microscopy," New York, John Wiley & Sons, Inc., 1951.

Gage, S. H.: "The Microscope," 17th ed., Ithaca, N.Y., Comstock Publishing Associates, 1943.

CHAPTER 2 | The Use of the Microscope

The application of the principles given in the last chapter to the everyday use of the microscope is simple. It consists only in knowing what is the best possible performance that may be expected of any type of instrument and then setting it up so that it can give this performance. Broadly speaking, there are three types of microscopes to be found in biological and medical laboratories. These may be called, for purposes of convenience, the *freshman microscope,* the *medical microscope,* and the *research microscope.* Let us take each of these in turn.

THE FRESHMAN MICROSCOPE

Specifications. The freshman microscope, of which a typical example is shown in Fig. 2-1, is usually provided with the following features.

Stand. A horseshoe foot on which is mounted an inclinable body composed of three parts—the mirror holder, the stage, and the tube carrier—all rigidly connected. The plastic stage is furnished with two clips. The tube carrier is furnished with a rack-and-pinion coarse adjustment with the aid of which the tube may be moved up and down over a distance of between 2 and 3 in. There is also a fine adjustment which provides geared-down movement of the tube over a distance of from ⅛ to ¼ in.

Tube. The tube is of fixed length, usually 160 mm, and is of smaller diameter at the upper end to fit the ocular or eyepiece. The lower end carries a three-lens turret—still called a *triple nosepiece* by microscope

32   The Microscope

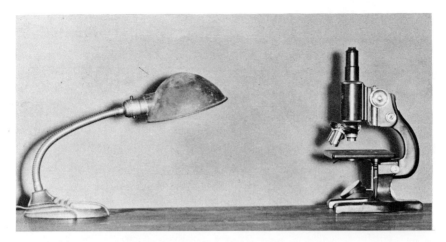

Fig. 2-1 | Freshman microscope with suitable illuminant. The microscope is an American Optical Company model No. 65, and the illuminant is a simple gooseneck lamp.

manufacturers—which, in cheap microscopes, usually rotates on a simple cone bearing.

Objectives. It is usual to provide microscopes of this type with ×3.5, ×10 (16-mm, N.A. 0.3), and ×40 (4-mm, N.A. 0.6) achromatic objectives.

Ocular. One ×10 ocular is all that is desirable.

Illuminating System. No special lamp is required, and it is not necessary to have a flat, as well as a concave, side to the mirror.

Setting Up the Freshman Microscope. This type of equipment is perfectly adequate for teaching elementary classes in biology. It is as simple to set up as it is to use, and every beginning student should be drilled in the following routine, illustrated in Figs. 2-2 to 2-7:

1. (Fig. 2-2.) Tilt the body back to a convenient angle, turn the ×10 lens into place, and adjust the concave side of the mirror until maximum illumination is obtained (Fig. 2-3).
2. (Fig. 2-4.) Place a slide on the stage under the clips.
3. (Fig. 2-5.) Use the coarse adjustment to bring the object into focus.

Setting up the high-power (×40) lens should never be attempted until the student is well practiced in steps 1 to 3. Then:

4. (Fig. 2-6.) After the object which it is desired to study under the high power has been located under the medium power, rack up the body and turn the high power into place. Then, looking in from the side as in the illustration, lower the high power with the coarse adjustment until it not quite touches the coverslip.
5. (Fig. 2-7.) While looking through the microscope, bring the lens up with the fine adjustment until the object comes into focus.

Figs. 2-2, 2-3, 2-4, and 2-5 | Setting up a freshman microscope. Fig. 2-2. The microscope is inclined to a suitable position. Fig. 2-3. The mirror is adjusted to throw light into the ×10 objective. Fig. 2-4. A slide is placed on the stage under the clips. Fig. 2-5. The tube is racked down with the coarse adjustment until the object comes in focus.

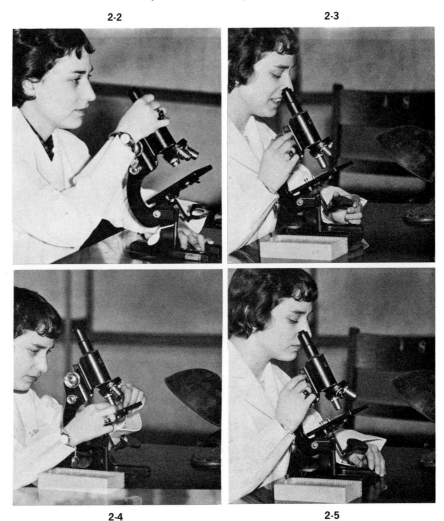

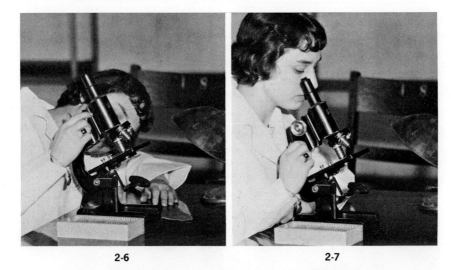

2-6　　　　　　　　　　　2-7

Figs. 2-6 and 2-7 | Setting up the high power on a freshman microscope. Fig. 2-6. After the object has been centered with the ×10 objective, the ×40 is swung into position and lowered, while being examined from the side, until it not quite touches the slide. Fig. 2-7. The tube is racked up with the fine adjustment until the object is in focus.

All this is so simple that only two things can go wrong, and these are easy to correct:

1. The fine adjustment jams before the object comes into focus. Correct by racking the tube up until the lens is well clear of the slide. Spin the fine adjustment down (i.e., away from you) until it reaches the end of its travel. Then repeat steps 4 and 5.
2. There is insufficient light with the high power even though the image is sharp. This is caused by either of two things:
    a. The flat, instead of the concave, side of the mirror has been used.
    b. The light source, such as the gooseneck lamp in the illustration, is too close to (*not* too far from) the mirror. The mirror acts as a lens (refer back to Fig. 1-9). The nearer the light source is to the mirror, the larger will be the image and the more diffuse the light. The further away, within reason, the light source, the smaller will be the image and the more concentrated the light.

THE MEDICAL MICROSCOPE

Specifications. Two types of medical school microscope are currently in use: those with a separate illuminant, such as the one shown in Fig. 2-8, and those with built-in illumination, such as the one shown in Fig.

2-18. The latter are very much the best, though they are still considerably more expensive and therefore not in universal employment. Apart from the illuminating system, the specifications given below apply to both types of microscope.

Stand. A horseshoe foot on which is mounted an inclinable body composed of four parts: the mirror holder, the substage, the stage, and the tube carrier. The mirror mount is the same as that on a freshman microscope. The substage is a ring into which the condenser fits. This ring is provided with rack-and-pinion focusing. The stage is usually furnished with a detachable mechanical stage, a device for providing rack-and-pinion controlled movement of the slide along rectangular coordinates. The tube carrier has the same coarse and fine adjustments as the freshman microscope.

Tube. Most of these microscopes can be obtained either with a monocular tube, of the type shown, or with a binocular tube of the type shown on the research scope discussed in the next section. The monocular tube is perfectly adequate for class instruction, but for routine work or prolonged observation the binocular tube saves so much eyestrain that it is well worth the extra money. Medical students buying their own microscopes are strongly advised to obtain one with interchangeable monocular and binocular bodies, even though they can only afford, for the time being, to purchase the monocular one. There is little difference in cost, and the subsequent purchase of the binocular body will adapt the instrument to the office routine of a practicing physician or his technician. The tube is furnished with a three-lens turret—four-lens turrets are better left to the research microscope, under which they will be discussed.

Objectives. Microscopes of this type are best provided with $\times 10$ (16-mm, N.A. 0.3) and $\times 40$ (4-mm, N.A. 0.65) dry objectives and with a $\times 90$ (2-mm, N.A. 1.2) oil immersion. Achromatic objectives are perfectly adequate for any type of use to which medical microscopes can legitimately be put. Medical students sometimes buy apochromatic objectives on the ground that since they are so expensive, they must be better; this same reasoning would justify harnessing a racehorse to a plow. Some manufacturers offer a $\times 40$ N.A. 0.75—or even 0.80—achromatic objective. This is a desirable, but not necessary, investment, and its short working distance is a great disadvantage. A four-lens turret with the addition of a $\times 3.5$ objective is a sheer waste of money because the condenser would have to be removed every time it was used. The proper fourth lens on a microscope of this kind is a $\times 20$ (8-mm, N.A.

0.75) achromat. These are expensive but extremely useful, particularly in histological work.

Oculars. Since the substage condenser permits the objectives to work at relatively high numerical apertures, a ×15 ocular is a useful adjunct to the standard ×10.

Substage Condenser. An ordinary Abbe condenser, with a maximum numerical aperture of about 0.9, is all that is wanted. It will, of course, have a built-in iris diaphragm, but neither the condenser nor the diaphragm should be in centerable mounts. These mounts require constant adjustments, which are necessary only when working with apochromatic objectives at the limit of their resolution.

Illuminating System. All this equipment is relatively worthless *unless it is accompanied by a proper lamp*. Those doubting this point should reread Chapter 1. A proper lamp is one having a small intense illuminant, a focusable condensing lens, a field iris in front of the lens, and a filter holder in front of the field iris. The intensity of illumination should be controllable, either through the provision of a rheostat or, in the writer's opinion better, through the purchase of a set of neutral density filters. The cost of this lamp should be carefully balanced against the increased cost of purchasing a microscope with an illuminating system built into the stand. Illuminating systems built into the substage are not very satisfactory, but those built into the stand permit the erection of Kohler illumination (described below) almost without effort. These stands are now available in a price range which competes with the combined cost of a regular microscope and the first-class lamp necessary to get the best from it. There are still, however, a considerable number of medical students who use separate lamps, and a description of setting up both types of microscope is therefore given.

The equipment described above is perfectly adequate for teaching histology, cytology, bacteriology, and any medical school course. It is of little value to elementary courses because of the difficulty involved in setting it up.

Setting Up a Medical Microscope with a Separate Illuminator.

A few words will have to be said on the practice of microscopy before giving a step-by-step description. Lenses are theoretically designed to work at their best when they are examining a self-luminous object. This theoretical ideal is closely approximated if an image of a self-luminous object is superimposed on the transparent object actually being examined.

Fig. 2-8 | A medical microscope with a suitable illuminant. The microscope is an American Optical Company model No. 35. The lamp is an American Optical Company model No. 735 and carries a neutral density filter.

This was actually possible in the early days when microscopists used oil lamps with broad wicks, and the original function of the substage condenser was to cast an image of this big flame into the field of the object. Some lamps had wicks as broad as 4 in. so that the image of the flame would fill the field of view of the objective. Images of the spirally coiled filaments of incandescent lamps are not, to put it mildly, so satisfactory.

The same optical conditions are met if the light radiating from a point source is gathered by a well-corrected lens and if the surface of the lens is used as the luminous source. There was a period in microscopy when everyone was trying to make a point source of light, but this is now obsolete.

The contemporary happy compromise is known as *Kohler illumination*. In this system light from a compact source—a lamp with small, closely coiled filaments—is gathered by a large lens that projects it as a fairly well-corrected beam. An iris diaphragm immediately in front of this lens is used as the actual light source, and an image of this iris is focused on the slide by the substage condenser.

Reference should now be made to Fig. 2-9, which shows, in diagrammatic form, the setup photographed in Fig. 2-8.

To produce Kohler illumination in practice, the student must be conscious of the necessity of forming two images. The first is an image of the lamp filament thrown by the field condenser roughly into the plane

38 The Microscope

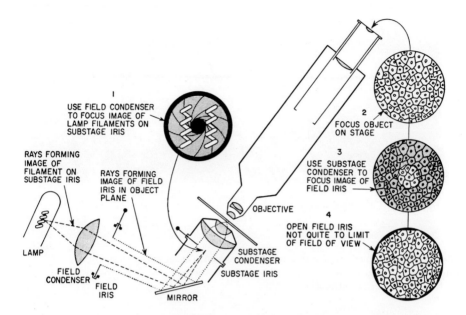

Fig. 2-9 | Diagram showing the four stages in setting up Kohler illumination.

of the substage iris. The second is an image of the field iris thrown exactly into the plane of the object by the substage condenser. These images are produced, and the system centered, by the following routine (Figs. 2-10 to 2-17):

1. (Fig. 2-10.) Swivel the turret to bring the ×10 objective into use. Turn on the lamp and adjust the mirror until enough light is obtained roughly to focus a slide on the stage.
2. (Fig. 2-11.) Open up both substage and field iris. Place a dark, neutral density filter in front of the lamp and focus the slide accurately.
3. (Fig. 2-12.) Shut down field iris and remove eyepiece. Look down the tube, with the eye about 2 in. away. Manipulate mirror, position of lamp, and tilting device of lamp until the lamp filaments are centered in the objective.
4. (Fig. 2-13.) Open field iris and lay a sheet of white paper on the mirror. Use the focusing device of the field condenser to throw an image of the lamp filament on the paper. A slight shift of the focusing device will then bring this image into focus on the substage iris [Fig. 2-9(1)].
5. (Figs. 2-14 and 2-15.) Close the field iris and focus an image of the field iris onto the object with the substage [Fig. 2-9(3)]. If the image is not central, adjust the mirror. Make sure that the object is still in focus. Open the field iris until it just delimits the field [Fig. 2-9(4)].

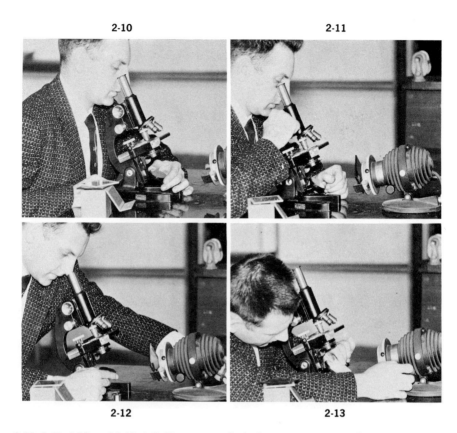

2-10 2-11

2-12 2-13

Figs. 2-10, 2-11, 2-12, and 2-13 | Setting up a medical microscope. Fig. 2-10. The mirror is moved until the ×10 objective is filled with light. Fig. 2-11. The neutral density filter is put in place and a slide focused with the coarse adjustment. Fig. 2-12. The eyepiece is removed, and the mirror and lamp are adjusted until the filaments are seen. Fig. 2-13. A piece of paper is held just in front of the mirror and an image of the filaments focused on this paper.

6. (Figs. 2-16 and 2-17.) Remove the eyepiece and look down the tube from about 4 in. away. Close the substage iris—which has been wide open since step 2—until it cuts off from one-tenth to one-fifth of the field. The objective will now be working at a desirable 80 to 90 per cent of its maximum aperture.
7. Shuffle the neutral density filters until the desired intensity of illumination is obtained.

Steps 1 to 4 do not have to be repeated until either the lamp or microscope is moved. Microscopes with an illuminant built into the stand never require these adjustments to be made. Steps 5, 6, and 7 must be

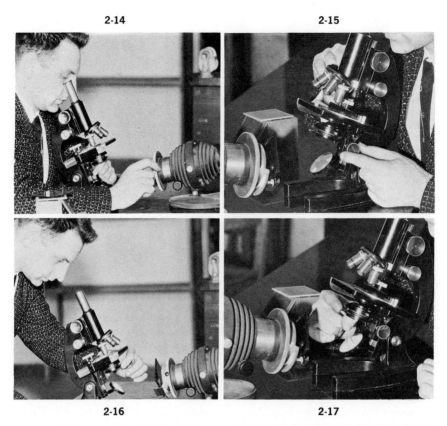

Figs. 2-14, 2-15, 2-16, and 2-17 | Setting up a medical microscope (continued).
Fig. 2-14. The field iris is closed. Fig. 2-15. The substage condenser is used to focus the image of the field iris in the same plane as the object on the stage. Figs. 2-16 and 2-17. The eyepiece is removed and the substage iris closed until it is seen just to cut off the edges of the back lens of the objective.

repeated every time either the objective or the slide is changed. Each objective works at a different N.A. and has a different field of view. The former must be adjusted with the substage iris and the latter with the field iris. No two slides are of the same thickness, so the substage condenser must be refocused for each slide.

The procedure for setting up the oil-immersion lens is much the same:

1. Repeat steps 1 to 5 with the ×10 objective.
2. Swing the ×40 objective into place and, if possible, find and center the field to be examined under the oil-immersion lens.

3. Rack up the tube and swing the ×90 objective into place. Place a small drop of immersion oil in the center of the field and, looking in sideways (compare Fig. 2-6), rack the tube down until the front of the objective just touches the oil.
4. Focus downward with the fine adjustment until the object is in focus. Shut the field iris until the field is delimited and, if necessary, sharpen the image of this iris with the substage condenser.
5. The substage iris is left wide open, so step 6 is unnecessary. The N.A. of the substage condenser, since it is working in air, cannot be greater than 1.0. The maximum N.A. of the objective is 1.2, so a working N.A. of 1.0 is just about right.

The only thing likely to go wrong in all this is in focusing the substage condenser. If the slide lifts when the condenser is brought up, before the field iris is in focus, there is nothing to be done. The slide is too thick, and it is necessary to be content with a second-class image.

Setting Up a Medical School Microscope with a Built-in Illuminator. Microscopes of this type are much easier to use and will undoubtedly replace, in time, microscopes with a separate lighting system. These microscopes have a lamp and a field condenser permanently set up to be in focus for color illumination, so that it is necessary only to center the substage and then to use the substage diaphragm to match the numerical aperture of the condenser to that of the objective. Both the lighting source and substage iris occasionally get out of center and are always provided with some device for centering them. The technique of centering is exactly the same in theory as for a separate lamp (as described above) but is much easier in practice. Assuming, however, that the lamp and the substage iris are centered, the following steps are necessary:

1. Focus a ×10 objective on a slide. Close the field iris (Fig. 2-18).
2. Use the substage focusing screws (Fig. 2-19) to focus an image of the field iris in the same plane as the object. Then open the field iris until it just fills the field.
3. Remove the eyepiece and look down the tube from about 1 ft away; then close the substage iris (Fig. 2-20) until it fills about two-thirds of the field. Replace the eyepiece, and the microscope is ready.

Many models of microscope, including the American Optical Microstar shown in Fig. 2-18, have a supplementary lens which is used to fill the field with light when using a ×3.5 objective. This is pushed in and out of position (Fig. 2-21) as required. Any difficulty encountered in setting up the microscope is almost invariably the result of forgetting to remove this supplementary lens. Even if the supplementary lens is to be used, it is necessary to get the substage in focus before doing so.

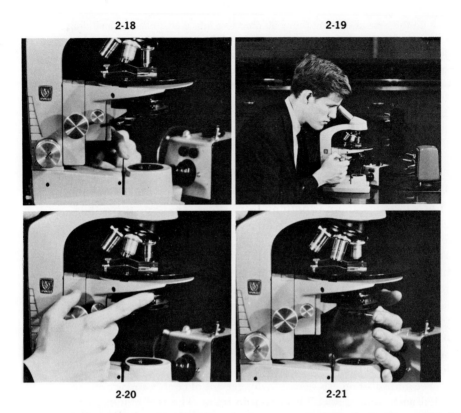

Figs. 2-18, 2-19, 2-20, and 2-21 | Setting up a medical school microscope with built-in illuminator. Fig. 2-18. The field iris is closed after the ×10 objective has been focused on an object. Fig. 2-19. The image of the field iris is focused with the aid of the substage condenser. Fig. 2-20. The N.A. of the substage condenser is matched to that of the objective with the substage iris. Fig. 2-21. Turning in the supplementary lens for use with a ×3.5 objective.

THE RESEARCH MICROSCOPE

Specifications. A description of a research microscope, of which a typical example with its illuminant is shown in Fig. 2-22, is included in this elementary book for two reasons. First, such microscopes are commonly set up for class demonstrations, and the student should certainly understand something about any instrument that he may look through. Second, there is a real justification for the use of these complex instruments even in elementary photomicrography, the subject of the next chapter. A first-class research microscope usually has most of the following features.

Fig. 2-22 | A research microscope and suitable illuminant. The microscope is an American Optical Company model No. 5, and the illuminant is an American Optical (formerly Silge and Kuhne) Ortho-Illuminator.

Stand. There are as many kinds as there are manufacturers of research microscopes. The stand carries the tube, the stage, the substage, and the mirror. Most contemporary microscopes have a fixed tube and are focused by moving the stage.

Tube. Research microscopes rarely have a monocular body, which was at one time required for photography. Most contemporary stands have an inclined binocular body (Fig. 2-22) when used exclusively for visual observation or a trinocular body (Figs. 2-27 and 3-19) when used for photography.

In most research microscopes the turret, which should carry four lenses, is attached to the body, not to the tube.

Stage. The stage of most research microscopes is both rotating and centerable. The first of these features is highly desirable since it permits the orientation of material both for observation and for photography in cameras which cannot be rotated. The second feature not only permits the stage to be centered but also permits fine adjustment of the rectangular rack-and-pinion movement of the built-in mechanical stage. It is quite common to have venier graduations for both the rectangular and rotating movements, but these are of more interest to geologists and metallographists than to biologists.

Substage. The substages of research microscopes are extremely complicated and consist of the following parts: (1) a device to permit substage condensers to be interchanged, centered, and focused; (2) an iris diaphragm, either mounted to permit lateral motion or supplemented by a separate lateral diaphragm; and (3) some fitting which will permit a supplementary lens to be placed under the condenser when low-power objectives are to be used. This is also found on some routine instruments (Fig. 2-21).

The two main methods of meeting these requirements are shown in Figs. 2-23 to 2-26. Figures 2-23 and 2-24 show the substage of a Bausch and Lomb DDE stand. The condenser, about to be inserted in Fig. 2-23, is mounted on a rectangular plate that slides in a dovetail slot. The iris diaphragm is mounted on a swing-out arm and has been swung out in Fig. 2-24 to show the rack-and-pinion that controls lateral motion. This rack-and-pinion is provided with a click stop to indicate the centered position, and the utmost care must be taken that the iris is clicked home before the substage is centered. The supplementary lens, also mounted on a swing-out arm, is shown out in both figures.

Figures 2-25 and 2-26 show the fork-type substage mount favored by the American Optical Company. The iris diaphragm is built into the condenser, and the whole unit is centered by the two screws shown in use in Fig. 2-25. This arrangement is less flexible, and also a great deal less trouble, than the mechanism preferred by Bausch and Lomb. The supplementary condenser lens, shown being inserted in Fig. 2-26, is mounted in a dovetail slide. This is a great deal more trouble to use than the swing-out lens on the Bausch and Lomb substage.

Both substage mounts are provided with rack-and-pinion focusing, and the particular American Optical model shown has, in addition, a fine adjustment. In Fig. 2-26 the coarse adjustment is the nearest milled head on the right of the figure. The fine adjustment is the small milled head in the left center of the figure.

Mirror. The only requirement of the mirror on a research microscope is that it be flat and easy to remove or turn aside. It is, of course, unnecessary with built-in, or substage, illuminators.

Objectives. The choice between apochromatic and achromatic objectives, the difference between which was explained in Chapter 1, depends on the use to which the microscope is to be put. Achromatic objectives have many advantages. They are cheap, accurately parfocal, of long

Figs. 2-23, 2-24, 2-25, and 2-26 | Substages. Figs. 2-23 and 2-24 illustrate the Abbe-type substage on a Bausch and Lomb model DDE. Fig. 2-23 shows the condenser being inserted on a dovetail slide. In Fig. 2-24 the substage iris has been swung out to show the rack-and-pinion lateral motion. In both Figs. 2-23 and 2-24 the supplementary condenser lens used with low powers has been swung out to the left. Figs. 2-25 and 2-26 show the fork type of substage mount on an American Optical Company model No. 5 microscope. In Fig. 2-25 the fingers are working the substage centering screws. The larger screws just above the hands are the stage centering focus. In Fig. 2-26 the supplementary condenser for use with low powers is being inserted in the dovetail slide. Notice the substage fine-adjustment control just below the left-hand condenser centering screw.

working distance, and of sufficiently high N.A. for most purposes. Apochromatic objectives have only the advantage of high resolution, more usually required for photographic than for visual work. It is necessary that all four lenses on the turret be in the same class since special (compensating) eyepieces are required for apochromatic objectives. The image given by an apochromatic objective with a regular eyepiece is far inferior to that given by an achromatic objective with the same eyepiece.

No matter what class of objective is chosen, the instrument should be provided with ×10, ×20, ×40, and ×90 (oil-immersion) objectives. A ×3.5 objective is even less practical on these instruments than on a medical microscope. In the apochromatic range there is often a choice between a ×90 N.A. 1.3 oil immersion and a ×95 (or ×100) N.A. 1.4 oil immersion. Only in the rarest circumstances does the 7 per cent increase in the theoretical resolution of the N.A. 1.4 justify the increased cost. Those who must have such a lens, because almost all of their work lies in the examination of objects at the extreme limit of the resolution of the optical microscope, would do well to substitute a ×80 (3-mm) oil immersion for the ×10 lens on the turret. These lenses have a field of view almost twice that of the N.A. 1.4 and are used to search the field for, and center, the object for subsequent examination under the high-resolution lens.

Oculars. These must be purchased as matched pairs. Although ×10 and ×15 are all that most people use, under ideal conditions of illuminating and centering the image made by an N.A. 1.4 apochromatic immersion lens will stand examination under a ×20 ocular. There is no point in using a ×20 ocular with lower-power objectives. A ×20 objective, for example, with a ×20 ocular gives a far less satisfactory image than a ×40 objective with a ×10 ocular.

Substage Condenser. This is just as important a part of the optical system as the objective. Nothing should be considered except an N.A. 1.4 achromatic condenser.

Illuminating System. The lamp described for use with the medical microscope (Fig. 2-8) can be used with a research microscope. It is not nearly so convenient, however, as the Ortho-Illuminator shown in Fig. 2-22 and in Figs. 2-27 to 2-32, which the author regards as the greatest advance in microscope illuminators of the last half century. This device permits automatic Kohler illumination with almost effortless ease in the control of centering and of light intensity. Many microscopes with illuminating systems built into the stand give effortless Kohler illumination and ease of centering, but the intensity can be controlled only by a rheostat—a situation wholly disastrous if color photography is to be attempted.

The Use of the Microscope 47

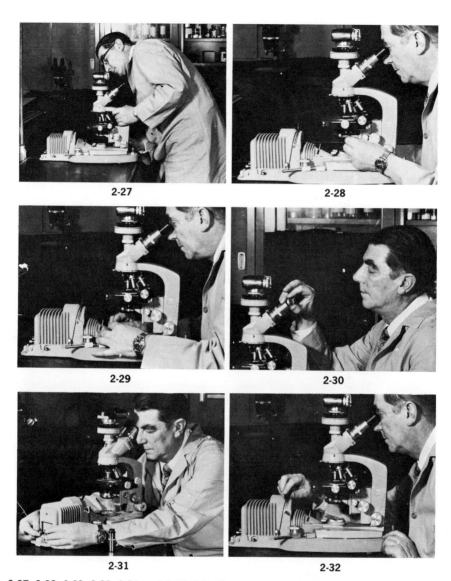

2-27

2-28

2-29

2-30

2-31

2-32

Figs. 2-27, 2-28, 2-29, 2-30, 2-31, and 2-32 | **Setting up a research microscope on an Ortho-Illuminator.** Fig. 2-27. The microscope is placed over the illuminator and moved until the ×10 objective is filled with light. Fig. 2-28. The substage iris is closed and the ×10 objective focused on the iris. The substage centering screws are then used to center this iris. Fig. 2-29. The substage iris is opened, the field iris is closed, and the ×10 objective is focused on the field iris. The field iris centering screws are then used to center this iris. Fig. 2-30. A pinhole eyepiece is substituted for one of the regular eyepieces. Fig. 2-31. The lamp centering screws are then used to bring the image of the filament, which is now seen through the pinhole, into the center of the field or view. Fig. 2-32. The intensity selector is moved until the desired intensity is obtained.

Setting Up a Research Microscope. There is no point in trying to use a research microscope that has not been properly adjusted. Many workers obtain worse images from these extremely expensive systems than can be obtained from a medical microscope simply because they do not know how to cope with the multiplicity of adjustments necessary. The two vital operations are centering the substage condenser and centering the light source. It is assumed that the reader is completely familiar with the operations described earlier in setting up a medical microscope.

Centering the Substage Condenser

1. Set the microscope approximately in place over the Ortho-Illuminator (Fig. 2-27) or set it up in front of a lamp as shown in Fig. 2-8.
2. Make sure, if the substage iris is of the type shown in Figs. 2-23 and 2-24, that it has been clicked into a central position. Then close the substage iris as far as possible and rotate the $\times 10$ objective into position.
3. Focus up and down, using, if necessary, both the coarse adjustment of the microscope and the rack of the substage, until the substage iris is in focus. Make sure by opening and closing it that you are looking at the substage iris, and not at some other iris in the system.
4. Use the condenser centering screws (Figs. 2-25 and 2-28) to bring the image of the iris into the center of the field.

The substage condenser is now centered with the objective system, provided the iris is properly centered with the condenser. Failure to find an image of the substage iris is sometimes caused by using too powerful an ocular, and a $\times 2$ or $\times 5$ ocular is an extremely useful accessory to have available if a $\times 10$ apochromatic objective is being employed.

Centering the Field Condenser

1. Open the substage iris and rack the condenser up to the top of its travel. Leave the $\times 10$ ocular in position.
2. Close the field iris (the lever seen in the dark slot just in front of the microscope on the Ortho-Illuminator) and use the coarse adjustment of the microscope to focus an image of the field iris. Make sure that you are actually looking at this iris by opening and shutting it.
3. Center the image of the field iris either
   a. On the Ortho-Illuminator by using (Fig. 2-29) the two screws lying just in front of the field iris, or
   b. On a detached lamp by adjusting with the mirror and, when necessary, moving the lamp.

The objective, the field iris (which is presumed to be centered on the field condenser), and the substage iris (which is presumed to be centered on the substage condenser) are now in line. It now remains to get the illuminant onto the same optical axis.

Centering the Light Source

1. Make sure that the image of the field iris is still in focus.
2. Replace one of the oculars with a pinhole ocular (Fig. 2-30). This device is most conveniently screwed to one of the clamp screws of the Ortho-Illuminator but must be purchased separately if another light source is used. On the Ortho-Illuminator the color selector must be set to zero and the intensity to maximum.
3. Open both the field iris and the substage iris.
4. Peep gingerly into the pinhole. If, as is almost certain, the glare is blinding, either slip a neutral density filter into the system or cut down the light emission with a rheostat.
5. The eye will now see a sharp image of the lamp filaments. Center these either
    a. On the Ortho-Illuminator by moving (Fig. 2-31) the two screws at the back of the lamp housing, or
    b. On a separate lamp by moving the lamp, tilting the lamp, and adjusting the mirror. This will probably throw the field iris out of line, as can instantly be seen by looking into the other barrel of the microscope, which still carries a normal ocular. It is now necessary, looking in each barrel alternately, to tilt the lamp, move the mirror, and move the lamp until both field iris and illuminant are centered. This infuriating procedure is what has caused most microscopists to abandon separate lamps.
6. Clamp the microscope in position on the stage of the Ortho-Illuminator. Those using a separate lamp will find it well worthwhile to provide themselves with a stage to which both microscope and lamp can be securely clamped.

Everything is now lined up, and it only remains to place a slide on the stage, set up Kohler illumination in the manner described for the medical microscope, and control intensity of illumination. On the Ortho-Illuminator (Fig. 2-32) this is done by rotating into position a disk of the required intensity.

All this laborious business, which stems from having a centerable substage condenser, is necessitated by the fact that the lenses in even the best turrets are not perfectly concentric. When, therefore, the $\times 20$, $\times 40$, or $\times 90$ lens is rotated into place and the substage refocused, it will be found that the image of the field iris is no longer perfectly central. It should be brought back to center by moving the substage with its centering screws. This movement will be of the slightest—if it is not, a complaint to the manufacturer should at once be made—and the slight off-centering of the field iris and light source will not be of importance. Never recenter the image by shifting the field iris, as it is of much more practical importance that the substage condenser be lined up with the objective than that the field condenser be lined up with the substage condenser. In fact, once the system has been clamped, no centering screws except those of the substage condenser should ever be touched unless the whole system is to be relined.

The ideal system, of course, is one in which each objective can be centered over the substage condenser after the latter has been lined up with the field condenser and the light source. Such systems exist—Figs. 1-15 to 1-17 could not have been taken without one—but they are altogether beyond the scope of this book. It might be added that they usually have to be lined up from scratch before each use, a procedure that takes about 3 hr.

Only a few operations with the research microscope remain to be described.

Setting Up an N.A. 1.4 Immersion System

The reader who followed the discussion in the last chapter will realize that this N.A. is obtainable only if the condenser, as well as the objective, is working in oil. Therefore:

1. Line everything up as previously described.
2. Set the slide on the stage and get in Kohler illumination with a ×40 objective. Find the area to be studied.
3. Remove the slide, rack up the tube, and place a drop of oil on the top lens of the condenser.
4. Refocus the slide with the ×40 objective and then refocus the substage condenser. Rack up the tube and make sure that the whole surface of the top lens of the condenser is joined to the slide with oil. It sometimes happens that there is too great a gap between the condenser and the slide. Remove the slide, lay a couple of coverslips on the oil on the condenser, add oil to the top of the coverslips, and replace the slide. This compensates, in effect, for having used too thin a slide. There is no cure for too thick a slide.
5. Place a drop of oil on the slide. Swing the N.A. 1.4 lens into position, lower it into contact with the oil, and then focus it.
6. It will usually be necessary, in practice, to shut down the substage iris slightly —to, say, N.A. 1.3—to reduce glare.

It may be necessary, after examining one field with this lens, to change slides and search the new slide with a ×10 lens to find the required field. The image of the field iris will of course be minute, and that is why the supplementary lens shown in Fig. 2-26 is provided. Swing or slide it into position, and the whole field of a ×10 lens will be illuminated. A similar lens is available for an Ortho-Illuminator.

Most N.A. 1.4 condensers will adequately fill the field of a ×20 objective. If much critical work is to be done with a ×10 objective it is best to have an N.A. 1.0 Abbe condenser, which can be exchanged for the N.A. 1.4. For occasional critical work, the top lens of the N.A. 1.4 condenser may be removed. The last operation to be described is centering the stage.

Centering the Mechanical Stage

This is usually necessary only when the microscope is to be used for photography.

1. Set up the ×10 objective.
2. Place a slide carrying crossed hairs on the stage and center it with the rectangular rack movement.
3. Unlock the rotating movement (Fig. 2-34). This extremely important step is often neglected, with consequent damage to the instrument.
4. Rotate the stage. The cross will describe the arc of a circle. Estimate the position of the center of this circle and try to bring this center into the center of the field with the stage centering screws (Fig. 2-33).
5. Use the rectangular rack movement to bring the crossed hairs back into the center of the field. Again rotate the stage. This time the crossed hairs will describe the arc of a much smaller circle.
6. Repeat steps 4 and 5 until the cross rotates on its own center.

Setting Up a Phase-contrast Microscope. A student with no previous experience in phase-contrast microscopy should read the account of this at the end of the last chapter. It is vitally necessary to get a phase contrast set up exactly right, and it is impossible to do this without some knowledge of the theory involved.

Figs. 2-33 and 2-34 | Centering the rotating stage.

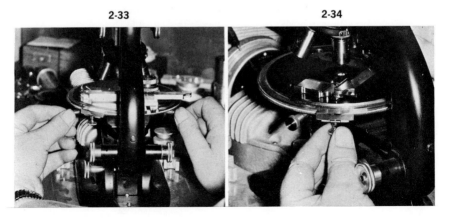

Each phase-contrast objective has to have a matched condenser since the annular disk underneath the condenser has to match the phase-contrast plate at the back of the objective. For this reason, the objectives and the condensers of the same manufacturer must always be used. Phase-contrast condensers, with their annular stops, are mounted in a turret that fits under the substage exactly as the objectives are mounted in a rotating turret on the front end of the microscope tube. In addition to the phase-contrast condensers, there must be a regular condenser, without an annular stop in it, which permits the whole substage system to be centered to the light. There are, therefore, two separate sets of substage centering screws. The regular substage centering screws, which in microscopes with interchangeable substage arrangements are used to center either the regular substage or the phase substage turret, are built into the microscope. The centering screws for the annular stops are built into the phase condenser turret itself in most microscopes, or else, as in the current American Optical models, they are activated by special probe handles which are inserted as required.

A convention which is at present accepted by most manufacturers of microscopes is that the regular condenser in a phase turret is labeled 0, the condenser matched to a $\times 10$ objective is labeled A, that matched to a $\times 20$ objective is labeled B, that matched to a $\times 40$ is labeled C, and that matched to the oil-immersion objective is labeled D. Compared with regular objectives, all phase-contrast objectives have very low numerical apertures, so that the phase-contrast condenser is never oiled to the slide. The apparatus is set up in the following steps.

Centering the Phase Substage Turret

1. In the unlikely event that the microscope being used for phase-contrast microscopy does not have a built-in lighting system, a light should be set up and centered with a regular condenser, using the technique described under Setting Up a Research Microscope earlier in this chapter.
2. Insert the phase turret in place of the regular condenser and rotate it until the zero condenser—that is, the one that has no annular stop—is in position. In almost all microscopes (Fig. 2-35), the number zero can be read only with the aid of a mirror.
3. Close the iris diaphragm of the phase substage turret and use the regular substage condenser centering screws to center the image of this iris through a $\times 10$ objective (Fig. 2-36).
4. Check that the light is centered by using a pinhole ocular in the manner described under Centering the Light Source earlier in this chapter.
5. Rotate the phase objective which is to be used into position on the nosepiece and use the mirror to observe the rotation of the substage phase turret until the

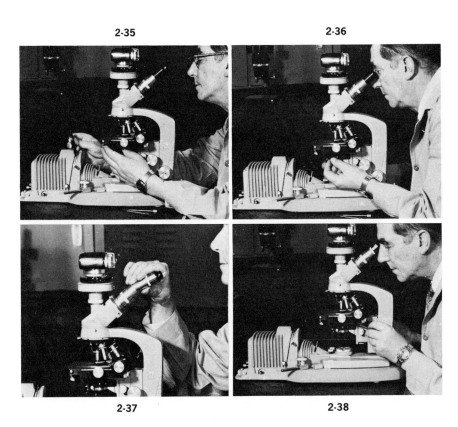

Figs. 2-35, 2-36, 2-37, and 2-38 | Setting up phase-contrast illumination. Fig. 2-35. The phase condenser turret is rotated until the notation 0 can be read in a mirror. Fig. 2-36. The image of the substage diaphragm is now used to center the substage condenser and to set up Kohler illumination. Fig. 2-37. A telescope ocular is placed in the tube and focused on the phase plate. Fig. 2-38. The annular stop is now centered against the phase plate.

condenser is matched to the objective. Insert a telescope eyepiece (Fig. 2-37), which comes with the phase equipment, and focus this on the annular phase plate on the back of the objective. This phase plate is seen as a dull ring of light. A much brighter ring of light, of exactly the same size, will also be seen. This is the image of the annular stop belonging to the condenser. If the two rings are not of exactly the same size, the wrong substage phase condenser has been rotated into position.

6. Use the annular stop centering screws (Fig. 2-38) to superimpose the two rings of light.
7. Remove the telescope and reinsert the regular eyepiece; the microscope is now ready for use.

Most people center all the phase plates against all the objectives when the microscope is first set up for phase-contrast work. This is less trouble than interrupting one's observations every time a new objective is swung into position.

SUGGESTED ADDITIONAL READING

Birchon, D.: "Optical Microscope Technique," London, Newnes Educational Publishing Co., Ltd., 1961.

CHAPTER 3 | Photomicrography

The easiest way to record a microscopic image is as a photograph. It must be emphasized that such records contribute little to understanding. The way to understand what is seen is to draw it, since this forces comprehension on the individual. The author cannot too strongly urge, on student and teacher alike, that freehand drawing should precede photography. Competent research workers use photography only to demonstrate to others what they themselves have understood. The essence of good photography is to know beforehand what the photograph should look like. The degree to which a photographer succeeds in demonstrating his point is a measure of his technical skill. A camera, particularly in photomicrography, cannot vouch for the truth of an observation. It can only, in extremely skilled hands, record what the observer believes to be the truth. In unskilled hands it cannot even do this, so that some knowledge of underlying theory is prerequisite to practice.

THE NATURE OF THE PHOTOGRAPHIC PROCESS

General Description. The recording of images on a photographic film is dependent on the instability of silver bromide. Crystals of this reagent sooner or later decompose to yield metallic silver. They decompose sooner if exposed to radiant energy from any source but have the peculiar, although not unique, property of being able to absorb such energy in amounts less than is required for spontaneous decomposition. Such crystals are naturally less stable than normal crystals and can therefore be decomposed by weaker reducing agents. In this lies the whole basic principle of photography. Sheets of crystals of silver bromide are exposed to light and dark. Those exposed to the light are rendered unstable and form a latent image on the sheet. This image is developed in a weak reducing agent which acts first on the unstable crystals in those areas to which light has penetrated. These crystals having been reduced to metallic silver, the image is fixed by dissolving out the more stable crystals, on which the developer has not yet had time to act.

The resultant image is thus blackened by the presence of silver in those areas on which light fell and is rendered transparent by the removal of silver bromide in those areas on which light did not fall. The result, were there only a single layer of crystals, would be a black-and-white image. In practice many layers of minute silver bromide crystals are supported in a gelatin film. In regions on which very little light fell, only the surface crystals are affected. The whole depth of the film, that is, all the crystals in the film, are affected in areas exposed to intense illumination. A thin layer of minute silver crystals appears very pale gray; a thick layer appears black. Hence the film is capable of recording a wide range of tones. It cannot do so automatically. It can only do so if it is exposed for just so long as will permit the faintest light to render the surface unstable and the strongest light just to penetrate to the depths. The actual intensity of the faintest and strongest is immaterial. The total exposure must be adjusted so that the faintest may have just time, and the strongest not too much time, to act.

Development is similarly not an all-or-nothing process. All silver bromide, exposed to reducing agents, will ultimately be reduced to silver. The purpose of photographic development is ideally to reduce only those crystals that light has rendered unstable; both time and temperature of development must therefore be rigorously controlled. Fixation, or the removal of undeveloped silver bromide, is a less exacting process, since metallic silver is only very slowly attacked by the silver bromide solvents employed. A final washing, to remove the fixing agent, leaves a relatively stable image composed of silver in gelatin.

The image so formed is, of course, a negative. That is, it is black where there was light and transparent where there was darkness. It is reversed to a positive, or print, simply by duplicating the process. The negative is laid against, or an image of the negative is projected onto, another layer of silver bromide in gelatin. This second layer is then, in its turn, developed and, provided that the exposure has been carefully controlled, will reproduce in whites, grays, and blacks all the shades of the original image.

There are, therefore, three steps in photography. First, the formation of an image on a layer of silver bromide for the necessary length of time. Second, the processing of this layer of silver bromide into a negative. Third, the reproduction of this negative as a positive, or print. The formation of an image through a microscope has been discussed in the last two chapters. Let us, before dealing specifically with the projection of this image onto negative material, discuss in more detail the types of this material that are available.

Negative Materials. The emulsion that is spread on a cellulose acetate base to make film contains, besides silver bromide and gelatin, additions designed to give each type of film its special characteristics. The characteristics of interest to the photomicrographer are resolution, speed, spectral sensitivity, and contrast. Many of these are cross-linked, so that it is, for example, impossible to obtain a film of both high speed and high resolution.

Resolution. The resolution of a film is just as important to a photomicrographer as the resolution of his lenses. It is obviously impossible to produce, as an image composed of grains of silver, pictures of particles of the same size as the grains. The grains of silver in the image can, in fact, be regarded as the pigment in a paint. When this pigment consists of extremely fine particles—as, for example, the carbon in the ink with which this page was printed—extremely fine lines can be sharply reproduced. If, on the contrary, the ink had been made from coarse, sooty particles, the outlines of the letters would inevitably have been blurred.

Speed. The speed of a film is the measure of its overall sensitivity to light. An image on a photographic film, just as much as an image on the retina of the eye, may be brilliantly or dimly illuminated. In either case there is a range of tones between the brightest and dimmest. A photographic film must be exposed just so long as will permit the brightest light in any particular image to penetrate the full depth of the emulsion while the dimmest light is just affecting the surface. Very sensitive, or fast, films require a short exposure, while very insensitive, or slow, films require a long exposure. The only reason for using fast films, which usually have relatively poor resolution, is to permit short exposures, and the usual purpose of a short exposure is to "stop" moving objects. Slow films of high resolution are therefore the only ones that need be considered for photomicrography.

Spectral Sensitivity. Different wavelengths of light have different energies. Short wavelength (blue) light beats like a staccato roll of drumsticks on the film. It exerts far more effect than the slow lapping of the long waves of red light. It follows that an ordinary photographic emulsion is very fast to blue light and very slow to red light. Films of this type are to all intents and purposes blind to red light and quite incapable of distinguishing between blue and white. A microscope slide stained in various shades of red and blue, photographed on such a film, would have clear gelatin where there should be all the shades of red and an almost uniformly massive deposit of silver where there should be all the shades of blue. A print from this negative would show the reds and pinks as solid black and all the blues as just a pale gray shadow. These films are obviously useless for photomicrography unless there is occasion to photograph a faint pink smear, the gradations of which are almost imperceptible to the eye.

Various additives to silver bromide emulsions cause them to become sensitized to red. These films are called *panchromatic*, often shortened to *pan*, since they can accurately reproduce gradations of all colors in gradations of gray. Bright red and bright blue will both be pale gray; dark red and dark blue will both be dark gray. It is easier to balance the films at the ends of the spectrum than in the middle, so that even the best of them see bright green as a darker shade of gray than bright blue or bright red. This can be compensated, if the exact rendering of shades of color as shades of gray is vital, by inserting a light-green filter somewhere in the optical system. This filter appears green because it absorbs part of the red and blue components of the light but transmits all the green. This intensifies the green components of the image, which thus are reproduced by the film in tones of gray equivalent to the blue and red components. Colored filters are also widely used in photomicrography deliberately to intensify one color at the expense of another. This will be discussed later.

Contrast. There is no pure black or pure white in the world. Both words are relative terms used to describe the opposite ends of a long scale of grays, each of which blends imperceptibly into the one next to it. Contrast is a measure of the length of this scale or of the number of steps of which it is composed. Imagine the scale of grays in a print to be a slope extending down from very bright to very dark. If this slope is very steep, with few steps, it can be descended in a series of harsh and memorable leaps. If the slope is gentle, with many steps, the descent is smoother and less dramatic. Exactly the same effect is produced by high- and low-contrast photographs. The high-contrast picture (Figs. 3-23 and 3-24) is dramatic and eye-catching—qualities not to be despised in illustrations of books and scientific articles. The low-contrast picture (Fig. 3-22) is a more truthful representation of the original. The photographer must decide for himself whether his picture should record an observation or tell a story.

Contrast may be controlled by the selection of negative material, in processing the negative material, and by the selection of the positive material on which the print is made. The last is the best. Extremely contrasty negative material should be used only to take pictures of objects, such as unstained cells, that lack contrast in themselves.

Selection of Negative Materials. It will by now have become apparent to the reader that the ideal film for general photomicrography is a high-resolution slow-speed panchromatic film of medium contrast. The three best known of the films are Kodak Plus X, Ansco Supreme, and Adox KB.

Each has its ardent supporters.

Positive Materials. Positive materials are also emulsions of gelatin and silver halides coated on a support. This support may be either paper, to produce a print, or glass, to produce a lantern slide. The criteria of positive materials are speed, contrast, and (in papers) surface.

Speed. All positive materials are relatively slow. The slowest are contact papers, coated with a silver chloride–silver bromide emulsion, which are pressed into contact with the negative before exposure. Contact prints are practically never used in scientific photography since projection printing, which may or may not involve enlargement, gives far better control of the process.

Prints are therefore made by projecting an image of the negative onto the face of bromide paper. There is little variation in speed between the products of various manufacturers, and paper should be selected entirely on the basis of contrast.

Contrast. The printing process is the right place to control the contrast of the final picture. Every attempt should be made to prepare a negative of medium density and medium contrast; from this, any kind of print can

Figs. 3-1, 3-2, and 3-3 | **The control of contrast by selection of printing papers.**
Fig. 3-2 shows a print of a normal negative on normal (No. 3) paper. Fig. 3-1 shows the same negative printed on very soft (No. 1) paper and Fig. 3-3 on very contrasty (No. 5) paper. The very contrasty print reproduces better, but the center print is better for observation.

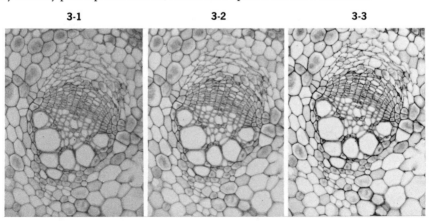

be produced. Some kind of print can be made from an extremely contrasty negative on soft paper; a print of sorts can be made from a flat negative on extremely harsh paper. Neither print will be as good as one made from a medium negative on medium paper. Most manufacturers make papers in five grades numerically designated from No. 1 (softest) to No. 5 (harshest). Figures 3-1 to 3-3 show the result of printing a good negative on paper Nos. 1, 3, and 5.

Surface. The only surface that can be considered for scientific prints is glossy. All other surfaces, made for pictorial effects, reduce resolution. The three best-known papers of this type are Eastman's Kodabromide F, Du Pont's R, and Ansco's Jet GL. The first is made in five, and the last two in four, contrasts.

PROCESSING PHOTOGRAPHIC MATERIALS

Everything so far written has concerned the materials in which a latent image is produced. This latent image is "processed" into an actual image.

The Darkroom. It is difficult to process negatives and impossible to process prints without an adequately equipped darkroom.

The room itself must be so constructed that it is capable of being made absolutely dark. Even the keyhole must be blocked and the base of the door fitted with a pad to prevent light seepage. On the other hand, the room must be light and well ventilated when not in use. Dark, damp cubbyholes are of no use as darkrooms because they encourage dirt and fungus, both equally ruinous to photographic equipment. The ceiling should be white, to reflect the light of the safelights used when processing papers. The walls should be medium to light green as a compromise between the cheerfulness of white and the risk of reflecting light escaping laterally from an enlarger. Concrete floors and unpainted wooden shelves are abominations that liberate dust and dirt to wreak havoc on the wet and sticky surfaces of newly processed material. A large household refrigerator should house films, papers, and reagents. Steel cabinets provide storage for everything else. An experienced photographer's darkroom, in fact, is a very different place from the dingy cell that architects of science buildings love to stick in an unwanted corner.

The basic furniture of a darkroom is a flat-bottomed stone or stainless-steel sink not more than 8 in. deep nor less than 4 ft long by 2 ft wide. Along the opposite wall, as far from the sink as possible, should be a sturdy bench. It is better to have two darkrooms: one "wet" for processing; the other "dry" for enlarging, loading dark slides, and the like.

There is no limit to the equipment that may be purchased for a darkroom. Minimum equipment includes safelights, developing dishes, developing tanks, and an enlarger. The nearest photographic dealer should be consulted, and his advice taken, by the inexperienced. Always buy the best possible enlarger, and particularly enlarging lens, that you can afford. There is no use going to great expense and trouble to produce a crisp, well-resolved negative and then projecting an image of this negative through a second-class lens. The enlarger should have a good condenser between the lens and the illuminant. Some enlargers have no condenser, which has the effect, desired by a few pictorial photographers, of softening the image.

Processing Negatives. Processing negatives consists of five steps. These are development, hardening, fixing, washing, and drying.

Development. The film to be processed may be in the form of either a roll or a sheet. The former is invariably, and the latter best, processed in a tank. Tanks for rolls of films consist of a lightproof case within which there is some device that allows the film to be spirally coiled without touching itself. Any dealer will demonstrate a variety of these tanks. Roll-film tanks have some kind of light-trapped aperture that permits the necessary reagents to be poured in and out in daylight so that the darkroom is required only for loading the tank.

Sheets of film are usually processed in rectangular open-top tanks and must therefore be kept in total darkness throughout the whole of development and hardening and the early stages of fixation. Films handled by this method are placed in rectangular metal frames, from the top of which horizontal arms project to rest on the sides of the developing tank.

The reagent used for developing the latent image is an alkaline reducing agent. The grain size of the film, on which resolution depends, can be increased during development by using solutions that are too strongly alkaline. Most fine-grain developers are solutions of p-diaminophenol, or some similar compound, buffered with borates in a solution of sodium sulfite that delays the spontaneous decomposition of the developer. The selection of the particular fluid to be used should be left strictly in the hands of the manufacturer of the film, who usually supplies either the formula or the reagent in ready-to-mix form. Experimenting with fancy developers is not recommended to the producer of technical photomicrographs.

Not only should the formula recommendations of the manufacturer be rigidly followed but also his specifications as to time and temperature. The photomicrographer who always uses the same film developed in the same developer for the same time at the same temperature can be certain that variations in his negatives are caused by faulty exposure, selection of the wrong filter, or imperfections in his optical system. None of these things can be corrected in the course of development; they can only be made worse. Without standardized development they are difficult to detect.

Most film-processing instructions offer times for tray development as well as tank. High-resolution photomicrographs are much better tank-developed. Tanks containing roll film, unless the manufacturer's directions read to the contrary, should be inverted about once a minute during the course of development. Cut film hanging in tanks should be removed once a minute and drained for 10 sec before being replaced in the developing solution.

Hardening. The process of development, after it has continued for the time specified by the manufacturer, should be abruptly terminated by transferring the film to a weak acid. Two per cent acetic acid is commonly used. It is much better to use a 3 per cent solution of sodium sulfate in 2 per cent acetic acid. This hardens the gelatin, which not only prevents mechanical damage but shortens the final drying time. Half a minute in plain acid or 3 min in hardener is standard.

Fixation. The image now consists of a mixture of silver and silver bromide. The silver bromide is dissolved in an acid solution of sodium thiosulfate, often miscalled *hypo*. A typical solution contains about 30 per cent sodium thiosulfate with 2 per cent sodium sulfite and 5 per cent sodium metabisulfite. Most people find it easier to buy the mixture ready prepared as acid-fixing salt.

The unwanted silver bromide is not directly soluble in the fixer. A complex double salt is first formed, and this is then dissolved out by the fixer. The first stage is ended when the film passes from a milky to a transparent appearance. The termination of the second, more important, stage cannot be seen. It is usual, therefore, to leave a film in fixer for about 10 min even though it appears clear after 3 or 4 min.

Washing. Every trace of sodium thiosulfate must be removed or the film will discolor rapidly. Twenty minutes in running water is sufficient for most films. If the film is washed in a tank, the stream of water must be taken through a tube to the bottom of the tank. Merely running a tap onto the surface is wholly inadequate.

Drying. All film should be rinsed in a weak wetting agent before being dried. This prevents water from gathering in droplets over the surface and leaving ring marks as it evaporates. Some wetting agents are damaging to photographic images, and it is therefore best to buy some compound, such as Photoflo, specifically marketed for the purpose.

Summary

Processing negatives consists, therefore, of the following steps:

Roll film (step 1 only in darkroom):

1. Load film into tank in darkroom (Fig. 3-4). Put lid on tank.
2. Pour in developer (Fig. 3-5), at temperature recommended by manufacturer, and leave for time recommended. Invert tank once a minute.
3. Pour developer back into bottle. Pour acid hardener into tank. Invert once or twice, then leave 3 min.
4. Pour off and discard acid hardener. Pour in acid fixer. Invert tank once or twice. Leave 10 min.

Figs. 3-4, 3-5, 3-6, and 3-7 | Developing 35-mm roll film in a tank. Fig. 3-4 shows the film being wrapped on the stainless-steel film holder of a Nikor tank. There are many other types of tanks. Fig. 3-5 shows the developer being poured in through the light-trapped lid. Fig. 3-7 shows the film being washed through a rubber tube running to the bottom of the tank. Fig. 3-6 shows the washed film being removed from the roll before being hung up to dry.

5. Pour fixer back in bottle. Remove cover from tank and flush out once or twice under tap. Either remove reel to large tank of running water or leave in small tank and wash through tube (Fig. 3-7) taken to bottom of tank.
6. Pour off water and rinse for 30 sec in diluted Photoflo or similar wetting agent.
7. Remove film from tank (Fig. 3-6) and hang up to dry.

Cut film (steps 1 to 4 in darkroom):

1. Place sheets in hangers (Fig. 3-8).
2. Set all hangers together in tank of developer (Fig. 3-9) at recommended temperature. Agitate for 15 sec. Then leave for recommended time. Lift hangers and drain for 10 sec, once every minute.
3. Transfer hangers to tank of acid hardener (Fig. 3-10). Agitate for 15 sec and leave 3 min.
4. Transfer hangers to tank of acid fixer. Agitate for 15 sec. and leave 10 min. Lights may be turned on after 2 min.
5. Transfer hangers to tank of running water (Fig. 3-11). Leave 15 min.
6. Dip hangers for 30 sec in diluted Photoflo or similar wetting agent and hang up to dry.

Processing Positive Materials. Processing papers is much less trouble than processing films both because the times are shorter and because the operations may be conducted in green light. The steps are just the same. Special developers are made for papers and are almost without exception solutions of *p*-diaminophenol and hydroquinone in a mixed solution of sodium sulfite and sodium carbonate. Neither the composition nor the temperature of paper developers is so critical as that of film developers. Most people keep one stock paper developer and use it at room temperature.

Development time varies with papers but is usually from 1½ to 3 min. Underdevelopment is a common fault. It is usual to stop development in a bath of 2 per cent acetic acid—hardening is not necessary—and fix in the same fluid used for films. Fixation and washing require a longer time for paper than for films since the reagents must be removed from the paper base as well as from the gelatin film.

Glossy prints are given an extra-high gloss and dried relatively flat by being rolled face down onto a highly polished chrome-plated sheet. This is usually in the form of a heated drum protected by a canvas belt, or a flat sheet with a canvas cover.

Papers are universally processed in trays, which should be considerably larger than the sheets used. The tray is filled out two-thirds full, and the paper, face up, is slid under the surface. The tray is continuously rocked during development, but prints are usually left in a large tray of fixer with only an occasional stir. The utmost care must be taken not to

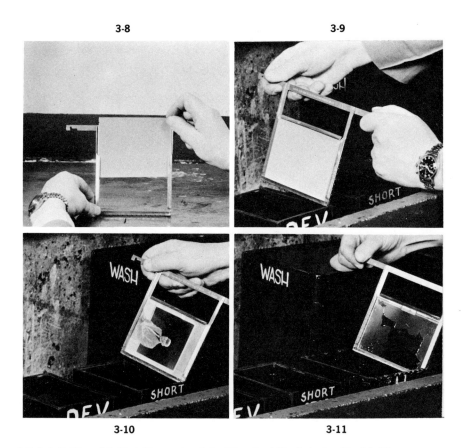

Figs. 3-8, 3-9, 3-10, and 3-11 | Processing 4- by 5-in. sheet film in tanks. Fig. 3-8. The film is inserted in a holder. Fig. 3-9. The film is lowered into a tank of developer. Fig. 3-10. The film is removed from developer and lowered into a tank of hardener or shortstop. Notice that the developed silver image is clearly visible in contrast to the white background of the undeveloped silver bromide. Fig. 3-11. After fixation the film is removed to the washing tank. Notice that the image is no longer visible because the silver bromide has been dissolved out in the fixer.

get fixer into the developing tray since even the slightest trace of contamination is ruinous to the image. If prints are handled with tweezers, a separate pair should be kept for each bath. If hands are used, the left hand alone should enter developer, and the right hand should enter fixer.

Prints can be properly washed only in a device that combines continuously agitated running water with periodic emptyings of the container. Many such devices are on the market. An hour is none too long to remove thiosulfate and silver from the paper base.

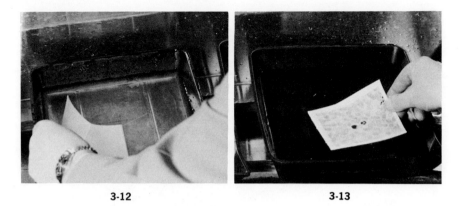

3-12   3-13

Figs. 3-12 and 3-13 | Processing paper. Fig. 3-12. The undeveloped paper is slid sideways under the surface of the developer. Fig. 3-13. The developed paper is slid sideways into fixer. Notice that the left hand is used for developer and the right hand for fixer. This prevents any possibility of contamination.

Summary

1. Slide exposed paper face up into developer in tray (Fig. 3-12). Rock continuously for 1 to 3 min.
2. Remove to acid stop bath for 30 sec.
3. Fix for 20 min in acid fixer (Fig. 3-13), rocking the tray at intervals. Do not allow prints to remain overlapped. Use large dish and few prints.
4. Wash 1 hr in print washer.
5. Roll wet print onto polished surface, cover with canvas, and leave to dry.

PHOTOMICROGRAPHIC CAMERAS

So far the discussion has concerned only the way in which photographic images are recorded. It is now necessary to turn to the apparatus with which these images are produced for record.

The first step is of course to set up a microscope to operate with the maximum possible efficiency. This has been fully described in Chapter 2, and no one should attempt to produce a photographic image until he can produce a perfect visual one. Given a perfect visual image, there are three ways of bringing this to a focus on a photographic film.

Direct-image Projection (Figs. 3-14 and 3-17). A vertical monocular microscope is set up, and a camera is lowered over the top. The front of the camera is fitted with a light-tight connector and the top of the

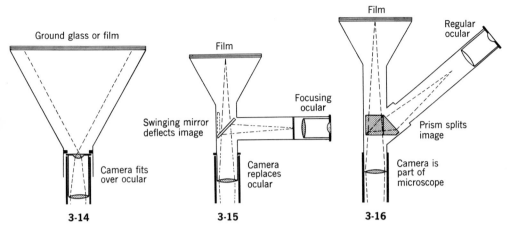

Figs. 3-14, 3-15, and 3-16 | **Three methods of projecting the image from a microscope onto a film.** Fig. 3-14 is the direct-image-projection method. The camera fits over the tube of the microscope, in which there is either a regular or a special photographic ocular. Fig. 3-15. The reflex-image-projection method. A mirror diverts the image into focus on a glass reticle seen on the right. A special eyepiece is focused on this reticle. The picture is taken by swinging the mirror on one side to project the prefocused image onto the film. These cameras fit into the top of the microscope in place of the eyepiece. Fig. 3-16. Split-beam method of projection. The camera and regular binocular observation eyepieces are both integral parts of the microscope. The beam is divided between these so that when the visual image is in focus the projected image is also in focus on the film.

camera with a groove into which may be fitted either a piece of ground glass or a film holder.

The advantages of this arrangement are in favor of quality rather than convenience. Particularly, it is possible to use special photographic oculars and, in the equipment shown in Fig. 3-17, to vary the distance of the film from the ocular. This permits the selection of specific areas of the field and also gives a flatter field than other methods. If the reader will reexamine Fig. 1-1, he will observe that the visual field is curved and that the rays leave the top lens of the ocular at a very wide angle. It follows that a flat film placed above a visual ocular will be out of focus at the edges and, unless it is placed very close, will show an excessively magnified image.

Every manufacturer of microscopic equipment offers a special photographic ocular in which Ramsden's disk is formed several inches above the top lens so that the image projected onto the film will create the impression of being the same size as that viewed in the eye. These oculars, moreover, are designed to project a flat, rather than a curved, image. The author prefers, of all the makes at present available, the series offered

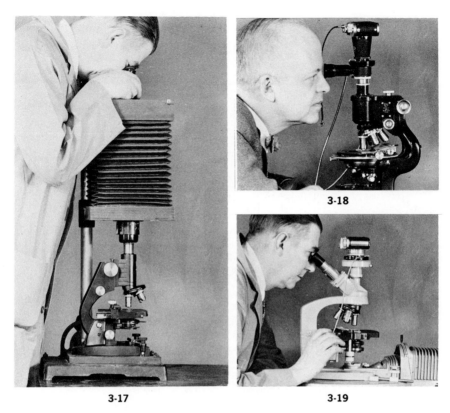

Figs. 3-17, 3-18, and 3-19 | Three cameras corresponding to the diagrams in Figs. 3-14, 3-15, and 3-16. Fig. 3-17 is a Bausch and Lomb model H camera corresponding to the diagram in Fig. 3-14. Fig. 3-18 is a Leica Ibso, corresponding to the diagram in Fig. 3-15, mounted on a Bausch and Lomb model DDE microscope. Fig. 3-19 shows an American Optical Company trinocular Microstar microscope and camera corresponding to the diagram in Fig. 3-16.

by Reichert. The photographic oculars of this company have the top lens fitted in a spiral focusing mount and may therefore be individually adjusted for the exact projection distance required.

The disadvantages of this type of equipment are very great. It is necessary to use a vertical monocular microscope. Then the camera must be swung into position and lowered into place. Next, final focusing must be carried out by observation of the image on a ground-glass plate; this is both difficult and inconvenient. Finally the ground glass must be removed and a film substituted. All this is so laborious and so time-consuming that, save for the most critical work, the disadvantages outweigh the advantages.

Reflex-image Projection (Figs. 3-15 and 3-18). A vertical monocular microscope is set up, and a roll-film camera with a built-in shutter is substituted for the eyepiece. The camera has a short horizontal tube into which the image can be deflected for focusing.

The advantages of this type of camera are that it is less trouble to set up than the type just discussed. The deflected image is arranged to come to a focus on a disk of glass bearing cross hairs. The focusing eyepiece is adjusted to give a sharp image of these hairs, and the fine adjustment of the microscope is then used to bring the image into focus; swinging the mirror aside will then cause a sharp image to be projected onto the film.

The disadvantages of this type of camera are in its lack of flexibility. The ocular is built into the tube of the camera and is so designed that the visual field of view will fill the whole area of the film, no matter what size the latter may be. The area to be photographed is therefore entirely dependent on the objective selected, and, usually, only the central portion of this area is in critical focus. It is also, compared with the next method, inconvenient to have to swing the mirror out of place before making the exposure.

Split-beam-image Projection (Figs. 3-16 and 3-19). A regular inclined binocular microscope has a beam-splitting prism built in over the objective. The rays that will form the image are thus split into two components. One portion passes directly upward toward the film, being interrupted only by a shutter to control the exposure. The other component is diverted toward the regular observational oculars.

The advantages of this system are enormous. To take a photograph, it is necessary only to activate the shutter. Nothing else need be moved, and regular visual observation need not, even momentarily, be interrupted.

The only disadvantages of this method are those of reflex cameras. That is, the lack of interchangeable oculars limits the flexibility of the system. Even this may be avoided, in the particular instrument shown, by substituting a monocular body for the film camera and rigging a direct camera on top of this.

TAKING A PHOTOMICROGRAPH

By now the reader should have a clear understanding of the principles of microscopy, the principles of photography, and the equipment available for photomicrography. With these in mind, there is little left to be discussed.

Selection of Film Size. The user has no choice of equipment used in the reflex and split-beam cameras just described. He is limited to the size, usually 35 mm, of the camera. This is, actually, no disadvantage, provided fine-grain film is used. Figure 3-20 shows a contact print and a ×12.5 enlargement from such a film. The granules in the bacteria are perfectly apparent in the print since the resolution of both the objective and the film was adequate. The writer has taken many photographs of such granules using 4- by 5-in. film and every artifice possible by the method of direct-image projection. The pictures so obtained are undoubtedly slightly better. It is doubtful whether they are sufficiently better to compensate for the inconvenience and effort involved. At lesser magnifications, 35-mm film is always adequate. Every photomicrograph in this book, except Figs. 1-15, 1-16, 1-17, E4-1, and E4-2, was taken with the equipment shown in Fig. 3-19.

Fig. 3-20 | Contact print and enlargement of 35-mm roll-film negative taken with the apparatus shown in Fig. 3-19. The little picture as reproduced is at a magnification of about 200. The large picture is at a magnification of about 1,000. The objects are the bacteria *Serratia marscens* stained with celestine blue B to show the nuclearlike granules.

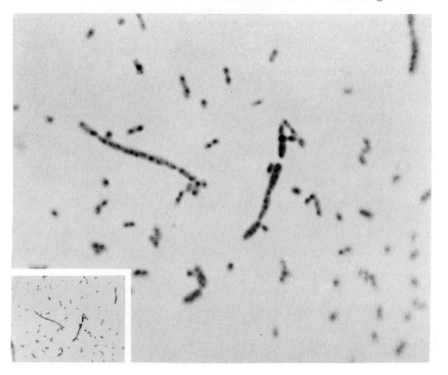

Fig. 3-21 | **Inserting cut film into a dark slide.** Notice the notches under the forefinger, which vouch for the fact that the film is right way around.

Cameras of the type used in direct-image projection usually take either 4- by 5-in. or 5- by 7-in. film. This film must be placed in a dark slide, and this, moreover, must be done in absolute darkness. For this reason, all cut film sheets are notched in one corner. The sensitive surface of the film is toward the user when the forefinger of the right hand (Fig. 3-21) rests on the notches. This being established, the film is slid into the grooves of the holder, and the dark slide is closed.

Glass plates coated with emulsions, which are really better in sizes of 5 by 7 in. and larger, cannot of course be notched. An experienced and sensitive finger can always tell which is the emulsion side. The beginner had best nibble the corner of the plate with moist lips. The emulsion adheres to the lips; the glass does not.

The dark slide (lower part of Fig. 3-21), which must be drawn out after the holder is in its grooves and before the exposure is made, has conventionally one side of its upper margin black and the other white. If the black is outermost when the film has not been exposed, and the white when it has, double exposures can be avoided.

Every film camera comes with directions for loading film, and there is no reason to repeat these here.

Selection of a Filter. One of the advantages of photographic over visual observation is that the film may be made to emphasize one or more features of a slide. This is done with the aid of colored glass filters inserted in the system. These filters remove from the illuminating system those colors which it is desired to emphasize, as dark shades, in the finished print. Let us take, as an example, the stem of Aristolochia shown in Figs. 3-22 to 3-24. In this, the nuclei and xylem are stained red with safranin. The cell walls are stained green. A well-balanced pan-

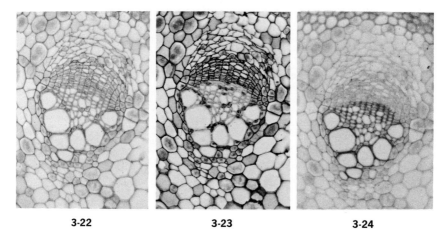

3-22 3-23 3-24

Figs. 3-22, 3-23, and 3-24 | The effect of color filters in photomicrography. The object photographed is a vascular bundle in the stem of Aristolochia stained by the method of Johansen described elsewhere in this book. The xylem and nuclei are scarlet. Other cell walls and cell contents are green. Fig. 3-22 shows a photograph on a properly balanced panchromatic film which renders the various colors in the proper intensity of gray. Fig. 3-24 shows the result of inserting a green filter into the system to emphasize the xylem. Fig. 3-23 shows the result of using a red filter, which emphasizes the cell walls and cell contents.

chromatic film will record the red and green as having equal values of gray. This is seen in Fig. 3-22. The insertion of a green filter into the system will remove red from the light beam so that the xylem, red in the slide, will appear black in the print (Fig. 3-24). This, although a conventional thing to do, is very doubtfully an improvement. A medium-red filter will remove green from the illuminant so that the cell walls (Fig. 3-23) will appear brilliantly defined. There is no limit to the number of filters that may be purchased, but a medium green (Wratten N or No. 61) and a medium red (Wratten A or No. 25) are all that were used in the preparation of illustrations for this book.

Exposure. No method more useful than trial and error has yet been developed for estimating the necessary exposure in photomicrography. Photoelectric exposure meters are intended to record the light reflected from a solid object to be photographed. A photomicrograph records the light transmitted through an object. Photoelectric meters adapted for photomicrography can do no more than record the total light emerging through the ocular. The proportion of this that is transmitted through the specimen depends on the thickness and method of staining of the latter.

Using a film-camera instrument, it is best to take a few frames at various exposures and to select the best. Figure 3-6, for example, shows the strip of film used in preparing Figs. E4-1 and E4-2. This is both inconvenient and expensive on larger films held in slide holders. In this case the dark slide should be manipulated so as to expose successively the whole, three-quarters, one-half, and one-quarter of the plate. Most fine-grain panchromatic films have a great deal of latitude—that is, they will produce a printable negative with a wide range of exposures—and most people keep exposures on test strips far too close together. An exponential series of three is perfectly satisfactory. If, therefore, experience suggests that 1 sec is the correct exposure, test pictures should be made at ⅓, 1, 3, and 9 sec. It is then easy by visual extrapolation to arrive at the right figure. A test strip at 1, 2, 3, and 4 sec would be relatively valueless.

There is no substitute for experience in judging which of the developed tests is correct. It is necessary, however, to remember that black and white are never present. The lightest area in the negative should therefore have some faint trace of gray. The darkest area should transmit some light.

Steps in Taking a Photomicrograph. If what has gone before has been understood, what follows will be easy:

1. Set up microscope (Chapter 2) to give the best possible visual image.
2. If necessary insert into the system a color filter that will produce the required emphasis.
3. Transfer the image, by direct projection, reflex, or split beam, onto a photographic film.
4. Prepare a test strip and develop for the time, at the temperature, and in the solution recommended by the manufacturer.
5. On the basis of step 4, make the final picture.

All this is just as easy as it sounds. Photomicrography is only the mutual application of a sound knowledge of photography to a sound knowledge of microscopy. With these prerequisites, it is easy; without them it is impossible.

SUGGESTED ADDITIONAL READING

Allen, R. M.: "Photomicrography," 2d ed., Princeton, N.J., D. Van Nostrand Company, Inc., 1958.
"Photography through the Microscope," Rochester, N.Y., Eastman Kodak, 1952.

PART TWO | The Preparation of Microscope Slides

CHAPTER 4 | Types of Microscope Slides

The equipment described is used to make four main types of microscope slides. These are wholemounts, smears, squashes, and sections. Examples of each are given in detail in Part Three, and it is intended here only to provide a brief summary to which the beginning student may refer.

Wholemounts. These, as the name indicates, are mounts of whole objects so preserved that the structure can be studied. Such objects must first be killed and fixed to preserve them in the shape they had in life. Contractile animals must be narcotized, even before this is done. The fixed material must then be washed, to rid it of the reagents used in fixing, and may be preserved in 70 per cent alcohol. Material of this type is opaque, so that, viewed under the microscope, it will appear only as a silhouette. It may be rendered transparent by soaking it in oil, a process which must naturally be preceded by the removal of water. This is accomplished by soaking the objects in alcohol. The "oil-cleared" specimens may be examined in this condition or, more usually, impregnated with a resin that preserves them as an insect is preserved in amber. Many things are by this process rendered too transparent for convenient study, so that they are, before clearing and mounting, stained in dyes, the better to show their internal structure. A few organisms are so tough that they may be placed directly in water-soluble gums for permanent preservation. This process of wholemounting is shown in diagrammatic form below.

78 The Preparation of Microscope Slides

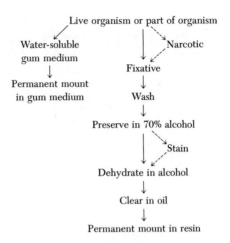

Smears. Smears, as the name indicates, are prepared by smearing a fluid such as blood, or a bacterial culture, as a thin layer on a slide. This layer is then stained to disclose its structure and air-dried. It may subsequently be preserved in dry form or mounted in resin. In diagrammatic form:

Squashes. Organisms obviously should not be squashed if it is desired to preserve the structure. The process is, however, invaluable when it is desired to study the contents of a cell without regard to the shape or relationships of the cell itself. In practice this method is usually confined to the study of chromosomes. In diagrammatic form:

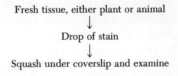

Sections. Organisms too large to be made into wholemounts, but too small to be studied by dissection, can conveniently be examined in the form of thin slices, or sections. This is, moreover, the only way in which the individual cells of organs can conveniently be displayed. Many plant and most animal tissues are too soft to be cut into thin slices and must therefore first be hardened. Reagents which fix the tissues in much the same form as they had in life can be selected for this purpose. After fixation the hardened tissues must be washed, and they are then commonly preserved in 70 per cent alcohol. Most tissues, even after hardening, must be supported during the process of cutting and to this end are conventionally impregnated with wax. This process, just as the impregnation of wholemounts with resin, involves preliminary dehydration and clearing.

When speed is of more importance than quality, the fresh tissue may be hardened by freezing and cut in this condition.

The sections, however cut, may be considered as objects to be made into wholemounts. That is, they must, in their turn, be stained, dehydrated, cleared, and mounted in resin. Even before this, sections cut in wax must have the wax removed and be rehydrated. In diagrammatic form:

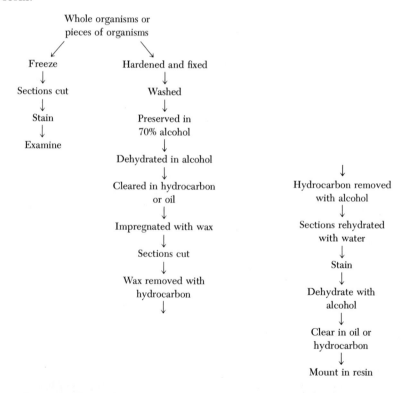

These simple outlines must now be expanded by a consideration of the various reagents used before being illustrated by specific examples.

CHAPTER 5 | Materials and Equipment

This brief introductory chapter is intended only for those who have never made a microscope slide and so are unacquainted with any of the names of the pieces of equipment used in its production. As these names will be used freely throughout the other chapters in this book, it is well here, by reference to illustration and descriptions, to make sure that the beginning student knows what the terms mean.

Slides and Coverslips. The microscope mount consists essentially of something intended for examination under the microscope, which is held between a slide and coverslip. The slide is almost invariably a piece of thin glass 3 by 1 in., although on rare occasions big series of sections are mounted on 3- by 1½-in. or even 3- by 2-in. slides. It is a great mistake to use slides that are too thin, for not only are they very easily broken, but they also do not work well with the condensers of most modern microscopes. A thickness of approximately 1 mm is the optimum, but slides as thick as 1¼ mm are perfectly satisfactory. It is essential that the surface of the slide be flat and that the glass of which it is composed be as stable as possible, for it will have to be passed through a great number of reagents. These stable glass slides are usually called *noncorrosive*, and it is a great pity to use any other kind.

Coverslips are circles, squares, or rectangles of thin glass; they are rarely, particularly in the larger sizes, completely flat. They are made in four thicknesses known as No. 0, No. 1, No. 2, and No. 3, each of which has its own special use. No. 0 coverslips, which are about 0.09 mm thick, are very difficult to handle and should be used only on preparations that are to be examined with an oil-immersion objective. They are rarely used today since most people, even for immersion lenses, prefer a No. 1 coverslip, which averages about 0.15 mm in thickness. No. 1 coverslips are easier to handle and to clean than No. 0. No. 2 coverslips, which average about 0.20 mm in thickness, are generally used on wholemounts, which are not customarily examined with the highest powers of the microscope. They are sufficiently thick to be easy to clean without breakage and sufficiently thin to be used with anything except an oil-immersion objective. No. 3 coverslips, averaging about 0.30 to 0.35 mm in thickness, are used only in making covers for dry wholemounts, which are not described in this book. They should never be used for any other purpose.

Circular coverslips are available in sizes from ⅜ to ⅞ in. in diameter, and the size most commonly employed is a ¾ in. (18 mm). For class purposes a ½-in. size is much more convenient and is large enough for the majority of wholemounts. Square coverslips come in the same sizes as the circular ones. The choice between circular and square depends entirely on the preference of the mounter. Square coverslips are a trifle easier to handle, but round coverslips have the advantage that one can "ring" the slide if one is making fluid or dry mounts. These types of mounts, however, are not described in this book, for they are unsuitable for elementary students. Rectangular coverslips intended for use on 3- by 1-in. slides are always 22 mm across their narrowest dimension and may be obtained 30, 40, or 50 mm in length. It is rarely wise to use a coverslip longer than 50 mm, for insufficient space will be left on the end of the slide for the application of the label. Coverslips intended for use with 1½-in. slides are usually 35 mm on their shortest dimension, and those intended for use with 2-in. slides are generally 43 mm on their shortest dimension. Coverslips for use with larger slides are almost invariably 50 mm long.

Containers for Handling Objects. Objects intended for microscopic examination have to go through a variety of processes, which are described in the next few chapters. Either the object may be put through these processes and subsequently mounted or it may be attached to a slide and the slide put through the processes. In either case a special container is required. Small objects are usually transferred between solutions with either a pipette (Fig. 5-1) or a section lifter (Fig. 5-2), the choice between the two depending upon the size and shape of the object. Where the object will stay for only a short time in a solution, it is customary to employ syracuse watch glasses (Fig. 5-3). These are called *syracuse watch glasses* because they have replaced the more conventional type of watch glass in laboratory technique. They have the advantage over the conventional type of being less easy to upset and are so shaped that they may be stacked one on top of the other, both for storage and

Fig. 5-1 | Pipette.

Fig. 5-2 | Section lifter.

Fig. 5-3 | Syracuse watch glass.

for prevention of too rapid evaporation of the contained solutions. They would be far better called *syracuse dishes*, but the name watch glass is still in customary use. Embryological watch glasses (Fig. 5-4) are used less for soaking objects in various solutions than for embedding objects in paraffin. They are very convenient for the latter purpose. They may also be used in place of syracuse watch glasses and have the advantage that they are provided with a cover that prevents evaporation of the contained solutions. They are not, however, so convenient as the conventional stender dish (Fig. 5-5), which is furnished with a round ground-on cover. If this cover is touched with a little petrolatum, it is possible to keep alcohol for several days in the dish without loss. When an object has to be stored for any length of time in a fluid, it is much better to use a vial (Fig. 5-6), the type shown having a screw cap of plastic. These are more expensive than the old-fashioned corked vials, but they are so much

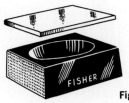

Fig. 5-4 | Embryological watch glass.

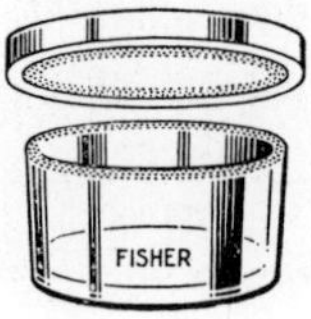

Fig. 5-5 | Stender dish.

Fig. 5-6 | Screw-cap vial.

Materials and Equipment 83

Fig. 5-7 | Standard coplin jar.

Fig. 5-8 | Rectangular slide jar.

better that they are worth the additional expense. They should be available to the mounter in a great variety of sizes. Where a number of very small objects is to be stored, it is convenient to place each object in a little vial plugged with cotton and then to accumulate these vials in a larger screw-cap vial containing the preservative fluid. In this way, many hundreds of minute specimens may be kept separate and safe for long periods.

Equipment for Handling Slides. Where the object to be treated is attached to a slide, it is necessary to have special equipment to keep the slides separate as they are put through the solutions. If only a single slide is being handled, it is possible to use a vial, provided that the vial is more than 1 in. in diameter. Almost invariably, however, several slides are handled at one time, and for this purpose it is almost universal to use a coplin jar (Fig. 5-7). These rectangular jars are furnished with a series of grooves that hold a number of slides apart. Most coplin jars hold six slides, but it is possible to obtain them to hold as many as twelve. The number of slides may be doubled by placing two slides back to back and sliding this sandwich into a groove, but this is very unsatisfactory because reagents diffuse so slowly from between the slides.

Coplin jars are made to handle only 3- by 1-in. slides. When one is dealing with the larger size, it is necessary to use a rectangular jar (Fig. 5-8) into which the slides are placed with their long edges downward. The jar shown contains a removable glass rack, so that a number of slides may be transferred from one jar to another without handling each one separately. This type is a great deal more expensive than the conventional rectangular jar with grooves down the side but is so much more convenient to use that it should be obtained where possible.

A set of eight jars, either coplin or of the rectangular type, is the minimum required for ordinary purposes of slide making. Coplin jars made of polyethylene are also available. Those with screw caps are particularly useful for the preservation of alcohols. Polyethylene jars are not satisfactory for staining solutions since many stains are absorbed by them.

In addition to the items mentioned, there are a large number of specialized pieces of equipment, such as microtomes, warming tables, paraffin, and embedding ovens, which are part of the regular laboratory equipment. These will be discussed in those parts of the book in which special preparations are described. The student will also need an ordinary dissecting set, but it is to be presumed that he will have this from previous courses in the biological sciences which he may have taken. If it is the intention of the instructor that the class prepare hand sections, it will be necessary to have section-cutting razors, either available for issue or made part of the student's set of instruments.

CHAPTER 6 | Fixation and Fixatives

There are a few cases where a living animal may be mounted directly on a microscope slide by placing it in a drop of a mountant and putting a coverslip on top. This, however, is the exception. It is usually necessary that living forms, or parts of them, be fixed in such a manner as to preserve them in the shape they had during life and hardened in order to render them of a consistency suitable to subsequent manipulations. Fixing and hardening agents are usually combined into one solution known as a *fixative*, and, from the practical point of view, the worker may require one or more of three functions from the solution employed. These functions are listed below.

PURPOSE OF FIXATION

1. To preserve the material in the shape it had before fixation
2. To preserve the nuclear elements of the material
3. To preserve the cytological elements of the material

All these functions can rarely, if ever, be secured from one solution, and each will be discussed separately.

Preservation of External Form. The loss of external form on the part of fixed material is brought about either by the contraction of the animal or by unequal diffusion leading to the distortion of cavities.

The contraction of the animal in many cases may be prevented by preliminary narcotization, which is often essential in the case of invertebrates. Moreover, in such animals as the Rotifera and Bryozoa, the fluid employed must contain an immobilizing agent if the external form is to be preserved successfully.

There appear today to be only three immobilizing agents of general value—a temperature between 60 and 75°C, osmic acid, and, to a far less extent, mixtures of acetic acid with chromic acid (chromic oxide) and picric acid (trinitrophenol).

The first of these agents—heat—obviously may be added to any known fixative. In the majority of cases, both cytological detail and histological detail are ruined by its use, yet it remains of great value for many of the marine Hydrozoa which are subsequently intended to serve as wholemounts or museum preparations. Osmic acid is unquestionably the most useful immobilizing agent that has yet been discovered, in that cytological, but not nuclear, detail is well preserved by its use. Many of the lower invertebrates retain far more of their original transparency with this than with any other fixative. However, it is both expensive to buy and dangerous to use, so that it cannot be recommended to an elementary class.

It is vertebrate embryological material that suffers most from distortion, and the preservation of its external form is best accomplished by solutions containing formaldehyde, potassium dichromate, or potassium dichromate plus sodium sulfate. It must be made quite clear that the word distortion is used here to indicate a change that produces a definite change of shape. Uniform shrinking and swelling without an accompanying change of shape are perfectly distinct processes; potassium dichromate (*inter alia*) produces the former and formaldehyde the latter, so that combinations of these two or of formaldehyde with Müller's fixing fluid are indicated for the preservation of the external form of delicate mammalian tissues.

Preservation of Cellular Detail. Preservation of tissues for section cutting is rather different from preservation of animals for wholemounts. The primary purpose of a protoplasmic fixative is to stabilize the proteins, preferably in a form which will permit the rapid penetration through them of dehydrating, clearing, and embedding agents. The principal reagents used for this purpose are known as *coagulants* since they precipitate the protein in a clot of finely shredded strands. Mercuric chloride and chromic acid are typical of the reagents used for this purpose. Acetic acid, an almost universal ingredient of fixatives, also coagulates nucleoproteins in a form which is highly misleading. The network of fine strands which is universally pictured in textbooks of histology as typical of a nucleus bears no relation to the structure of the living nucleus but is a simple artifact produced by the reagent. Noncoagulant fixatives such as osmic acid, widely used in electron microscopy in neutral or weakly alkaline solution, fix the nucleus as a homogeneous whole, a fact discovered nearly half a century ago and universally rejected until the advent of the electron microscope.

Some fixative reagents, notably formaldehyde, picric acid, and mercuric chloride, actually form compounds with the protein. The nature of these compounds greatly affects their subsequent staining reactions, which is the reason that a specific fixative is often stated to be necessary before a specific type of staining. Most fixative mixtures contain coagulants, which render subsequent embedding easier, together with anticoagulants, which control both the texture and the staining reactions of the material.

FIXATIVE MIXTURES

Dichromate Fixatives. Potassium dichromate was one of the earliest substances employed for fixing. It is, however, a very poor protein coagulant. In its original use, in the solution of Müller, it was used in combination with sodium sulfate, which is an excellent protein precipitant. Most of the modern dichromate fixative solutions have been derived from Müller, although in most the addition of mercuric chloride has made the sodium sulfate unnecessary. One of the best known is:

Zenker's Fluid:

Water	100 ml
Potassium dichromate	2.5 g
Mercuric chloride	5 g
Sodium sulfate	1 g
Glacial acetic acid	5 ml

This is one of the most widely used of the general-purpose histological fixatives. It permits very brilliant afterstaining and is in almost universal use in pathological laboratories. Too much should not be made up at one time, for it is not very stable. If large quantities are to be prepared, it is desirable to omit the acetic acid until immediately before use. Pieces should be fixed for about 24 hr and then washed in running water overnight.

For the fixation of botanical specimens, it is customary to use less mercuric chloride. The following is a well-known mixture:

Lavdowsky's Fluid:

Water	100 ml
Potassium dichromate	5 g
Mercuric chloride	0.15 g
Glacial acetic acid	2 ml

The lower concentration of mercuric chloride permits specimens to be left in this for a much longer period without danger of becoming hardened.

Many people prefer to add formaldehyde to these mixtures. The best known of these fluids probably is:

Helly's Fluid:

Water	90 ml
Potassium dichromate	2.5 g
Mercuric chloride	5 g
Sodium sulfate	1 g
40% formaldehyde	10 ml

This fluid must be made up immediately before use, and fixation should take place in the dark since the presence of light greatly accelerates the reduction of the dichromate by the formaldehyde. It is also better to wash out the material in the dark in a weak (4 per cent) solution of formaldehyde rather than water. This solution is equally applicable to plant and animal tissues that are intended for subsequent sectioning. Some people prefer, particularly for animal tissues, to have acetic acid as well as formaldehyde in these mixtures. The result is often referred to as *Formol-Zenker,* although the best known of these mixtures actually is:

Heidenhain's Fluid:

Water	90 ml
Potassium dichromate	1.8 g
Mercuric chloride	4.5 g
Glacial acetic acid	4.5 ml
40% formaldehyde	10 ml

This fixative, like that of Helly, should be prepared immediately before use or at least should be prepared as two solutions, one containing the acetic acid and formaldehyde and the other the remaining ingredients. Fixation should also take place in the dark, and the tissues should be washed out in weak formaldehyde in the dark.

Chromic Acid Fixatives. Chromic acid (actually chromic oxide) is widely used in fixatives, usually with the addition of acetic acid. The best-known zoological fixative of this type probably is:

Lo Bianco's Fluid:

Water	100 ml
Chromic acid	1 g
Glacial acetic acid	5 ml

This fixative is particularly adapted for use with small invertebrates and was developed by Lo Bianco for the fixation of marine forms. The fixative should be freshly prepared before use, and the object left in it for about 30 min, in the case of an invertebrate larva, to as long as overnight in the case of medium-sized Polychaetes. After fixation, the object should be washed in running water until no further color comes away. Very much weaker solutions are usually preferred by botanists. The one customarily recommended is:

Gates's Fluid:

Water	100 ml
Chromic acid	0.7 g
Glacial acetic acid	0.5 ml

This fluid is excellent for the fixation of plant chromosomes in root tips, etc. Specimens should be left in it overnight and then washed out in running water.

The addition of formaldehyde to chrome-acetic mixtures is very common in botanical practice. These solutions are known as *Craf* fixatives, a popular example being:

Navashin's Fluid:

Water	75 ml
Chromic acid	0.8 g
Glacial acetic acid	20 ml
40% formaldehyde	5 ml

This must be made up immediately before use, or one may prepare it as two solutions, keeping the formaldehyde separate from the chromic acid.

Mercuric Fixatives. Mercuric chloride is often used as the main protein precipitant in a fixative, as well as in combination with dichromate. It has the very grave disadvantage that steel instruments are destroyed instantly on contact with the solutions, so that one must use glass or plastic in handling specimens. There are, however, a number of excellent formulas. One of the best general-purpose fixatives ever invented is:

Gilson's Fluid:

Water	88 ml
95% alcohol	10 ml
Mercuric chloride	2 g
Glacial acetic acid	0.4 ml
Nitric acid	1.8 ml

This is a magnificent fixative for zoological specimens. Objects may be left for months without undue hardening; small objects are adequately fixed after a few hours. Specimens should be thoroughly washed in 70 per cent alcohol. This fixative is to be recommended to the beginning student for general use.

It occasionally happens that the microscopist desires to fix something that is covered in a very hard shell and, therefore, requires a fixative in which other desirable qualities have to be sacrificed in favor of extremely rapid penetration. One of the best of these mixtures is:

Carnoy and Lebrun's Fluid:

Absolute alcohol	33 ml
Glacial acetic acid	33 ml
Chloroform	33 ml
Mercuric chloride	*to saturation* (about 25 g)

This fluid may be used equally well with a hard-shelled arthropod or with a hard-shelled seed and will penetrate rapidly enough to preserve the whole. It cannot be used satisfactorily with any object containing fat, which will be dissolved, and is usually employed only when the principal interest of the worker is in nuclear fixation.

Picric Acid Fixatives. Picric acid (actually trinitrophenol) has been widely used in the last quarter century as a component of fixatives. The best-known formula undoubtedly is:

Bouin's Fluid:

Saturated aqueous solution of picric acid	75 ml
40% formaldehyde	25 ml
Glacial acetic acid	5 ml

It is a great pity that the use of this fixative should have become so widespread, for its only advantage is that objects may be left in it for a long time without becoming unduly hard. It has the disadvantage that picric acid forms water-soluble compounds with many substances found in a cell, so that sections cut from materials fixed in Bouin's fluid frequently show large vacuoles. It is also very difficult to wash the fluid from the tissues; even small traces of picric acid interfere with staining.

Bouin's fluid was recommended by its inventor for the fixation of meiotic figures but has been replaced largely for this purpose by:

Allen's Fluid:

Water	75 ml
40% formaldehyde	15 ml
Glacial acetic acid	10 ml
Picric acid	1 g
Chromic acid	1 g
Urea	1 g

Small pieces of tissues should be fixed overnight and then washed in 70 per cent alcohol until no more color comes away.

Other Fixative Mixtures. A great variety of other materials have been recommended as fixatives from time to time, but only three are of sufficient interest to be worth repeating here. The first of these is:

Kolmer's Fluid:

Water	87 ml
Potassium dichromate	1.8 g
Uranyl acetate	0.75 g
40% formaldehyde	3.6 ml
Glacial acetic acid	9 ml
Trichloroacetic acid	4.8 ml

This very interesting fixative was developed originally for the fixation of whole eyes, but it may be used very profitably in any place in which nerve structures are to be examined subsequently. The salts of uranium are widely used in fixatives intended solely for the central nervous system, but this particular formula is also useful for general purposes. Fixation should take place overnight, and then the material should be washed in running water.

Another very little known but admirable fixative is:

Petrunkewitsch's Fluid:

Solution A	
Water	100 ml
Nitric acid	12 ml
Cupric nitrate	8 g
Solution B	
80% alcohol	100 ml
Phenol	4 g
Ethyl ether	6 ml

Note: One part of A is mixed with three parts of B immediately before use.

This is very nearly as good a general-purpose fixative as the mixture of Gilson and has the advantage over Gilson's that objects fixed in it can be handled with steel instruments before being washed. It is not desirable to leave specimens in it too long (certainly not more than 2 or 3 days), and the fixative should be washed out in 70 per cent alcohol.

The third fixative is:

Bles's Fluid:

70% alcohol	90 ml
40% formaldehyde	7 ml
Acetic acid	3 g

This is one of the innumerable fixatives known to botanists under the generic name FAA.

OPERATIONS ACCESSORY TO FIXATION

Removal of Fixatives. Under each of the formulas given above, it has been indicated in what fluid the specimen should be washed to remove the fixative. This simple washing, however, is not always sufficient in the case of fixatives containing mercuric chloride or picric acid. Mercuric chloride occasionally gives rise to long needlelike crystals in tissues. These may be prevented by soaking the specimen, after it has been washed in running water overnight, in:

Lugol's Iodine:

Mix 1 g of potassium iodide with 0.5 g of iodine. Add 2 or 3 ml of water and shake until dissolved. Then dilute to 50 ml. If diluted to 150 ml, this becomes Gram's iodine.

Note: Iodine is very soluble in strong solutions of potassium iodide and very insoluble in weak solutions.

After the pieces have been soaked in this solution overnight, they should be transferred to either 70 per cent or 90 per cent alcohol, in which they should remain until no further color comes away. Specimens so treated never show the fine needlelike crystals after mercuric fixation, but there is no reasonable explanation of why this should be so.

Tissues fixed in picric acid naturally are a bright yellow. All this bright yellow color cannot be removed, as some of it is due to the formation of complexes with the proteins. However, there are two methods by which much more of the yellow color can be removed than by washing in alcohol alone. The first is to add a few grains of lithium carbonate to the 70 per cent alcohol in which the specimen is being washed. The lithium salt appears to free some of the bound picric acid. Another method, which is more troublesome but much more satisfactory, is to transfer the specimen to:

Lenoir's Fluid:

Water	70 ml
95% alcohol	30 ml
Ammonium acetate	10 g

This liberates almost all the bound picric acid and is much the best aftertreatment for picric acid–stained materials that has appeared in the literature. It might be explained that the objection to the retention of bound picric acid is that it interferes seriously with some forms of staining.

Treating Hard Materials. It so happens that some materials, even after fixation, are so hard that they cannot possibly be cut into sections. This hardening is caused either by the presence of calcium in the form of bone or calcareous plates or by the presence of chitin. The removal of bony or calcareous material is not so easy as it sounds, for if one were merely to hang the material in an acid mixture there would be a great hydrolysis of the protein. Therefore, it is necessary to have in the solution, besides acid, something that will prevent hydrolysis and swelling. The most commonly used reagent for this purpose is phloroglucinol. The following is a typical formula:

Haug's Solution:

95% alcohol	70 ml
Water	30 ml
Phloroglucinol	1 g
Nitric acid	5 ml

The flask containing the reagents should be warmed very gently under a stream of warm water: heating over a flame is likely to cause an explosion.

There is not much use in transferring objects to a vial containing this fluid, for they will fall to the bottom and rapidly exhaust the acid around them. It is much better to hang them with a fine thread of silk from the top of the vial. The decalcification is complete if the object no longer feels hard when pricked with a pin. If there is no part of the object where one may without damage apply the pin, it is very easy to find out if decalcification is complete by having a dentist or physician observe the specimen on a fluorescent screen by X ray. As soon as decalcification is complete, the object should be washed in large volumes of 70 per cent alcohol. Some people claim that phloroglucinol interferes with afterstaining and prefer to restrain the swelling of the tissues with mercuric chloride. The best mixture of this type is:

McNamara, Murphy, and Gore's Solution:

Water	80 ml
95% alcohol	10 ml
40% formaldehyde	8 ml
Mercuric chloride	2 g
Trichloroacetic acid	6 g
Nitric acid	1 ml

This is used in exactly the same way as the solution of Haug, but one must, of course, be careful not to handle the object with steel implements because the mercuric chloride will destroy them.

No really satisfactory method for softening chitinous materials has yet been discovered, though the following is better than nothing:

Jurray's Mixture:

Chloral hydrate	50 g
Phenol	50 g

Insects or other chitinous forms are fixed in the fluid of Carnoy and Lebrun and are transferred without washing to Jurray's mixture, where they remain from 12 to 24 hr. Then this mixture is washed out in chloroform, and the objects are embedded in paraffin. The chitinous exoskeleton of arthropods can easily be softened for wholemounts by soaking the alcohol-hardened specimen in 10 per cent potassium hydroxide.

Regular woody tissues can be softened for sectioning by soaking small blocks of the wood in warm water or, in extreme cases, by boiling them for a few hours. Such blocks are best cut on a sliding microtome (see Fig. 12-18), though reasonably thick sections can be cut on a freezing microtome (see Fig. 12-41) provided that a method of clamping the unfrozen block in position can be developed.

Another useful technique with woody tissues is that of macerating small pieces to remove the vessels, fibers, and the like. One of the best methods is:

Harlow's Macerating Solutions:

Solution A
  Saturated aqueous solution of chlorine (chlorine water)

Solution B
  Water   100 ml
  Sodium sulfite   3 g

Small slivers of wood are placed in solution A for about 2 hr and then transferred to solution B which has been heated to about 90°C for 15 min. Put the wood back into solution A for another 2 hr and then back into solution B, repeating the cycle until the wood fails to turn red when placed in solution B. Simple shaking will now usually reduce the material to its constituent fibers, though in very obstinate cases it may be necessary to shake with glass beads.

Narcotization. There are many small invertebrates that cannot be made satisfactorily into microscope slides after the process of simple fixation. These forms, such as the majority of small hydroids and worms, are contractile, so that it is necessary to narcotize them before fixation if they are to resemble the living form after mounting. The whole subject of narcotization is very difficult, for it requires great skill to add slowly small quantities of the selected narcotic and then to fix the object at the exact moment that it is completely narcotized but before it is dead. Many marine forms, particularly sea anemones, may be narcotized by adding small quantities of a saturated solution of magnesium sulfate to the water in which they are expanded. There are, however, two mixtures that can be recommended for general purposes. The first of these is:

Hanley's Solution:

Water   90 ml
Ethyl Cellosolve   10 ml
Eucaine hydrochloride   0.3 g

This is an excellent narcotic for small and delicate forms, such as Rotifers and Bryozoa. About one drop per 10 ml should be added and well mixed with the water in which the creatures are living. After about 10 min, further quantities may be added and left until narcotization is complete.

A useful narcotic for less delicate forms is:

Gray's Mixture:

Grind 48 g of menthol in a mortar with 52 g of chloral hydrate.

This mutual solution of menthol and chloral hydrate is lighter than water. A few drops placed on the surface of the water containing the specimen dissolve slowly, allowing both constituents to act rapidly and safely. This reagent is recommended for Coelenterates of all kinds.

SUGGESTED ADDITIONAL READING

Baker, J. R.: "Principles of Biological Microtechnique," New York, John Wiley & Sons, Inc., 1958.

Gray, P.: "The Microtomist's Formulary and Guide," New York, McGraw-Hill Book Company, 1954.

CHAPTER 7 | Stains and Staining

PRINCIPLES OF STAINING

Many objects, after being impregnated with mounting media (Chapter 9), become so transparent that their structure cannot be observed under the microscope. To overcome this difficulty, such objects are commonly stained—that is, impregnated with a color—which renders them more visible, as was explained in Chapter 1. In the case of materials of uniform chemical composition—for example, a section of wood—it does not matter very much what dye is used. Most sections, however, contain an assortment of components that may be distinguished by their ability to retain dyes of contrasting colors. Thus a section of a plant stem may be so stained that contrasting colors are exhibited by the lignin, the cellulose, the protoplasm, and the nuclei. A section of animal tissue may be so treated that bone, cartilage, muscle, nerves, white connective tissue, and nuclei stand out in vivid color contrasts. These triple, and even quintuple, stains are rarely necessary for the everyday business of teaching and research. In plant anatomy it is usually sufficient to distinguish woody from nonwoody tissue. In animal histology it is enough to distinguish the nucleus clearly from the surrounding cytoplasm.

A dye, whether it is derived from natural sources or synthesized, owes its color to the presence of a chemical group known as a *chromophore*. Typical chromophores are the azo group, found in methyl orange—

Azo

Indamine

among many other dyes—and the indamine group, found in methylene blue and celestine blue B. The mere presence of a chromophore does not impart color but makes it possible for the color to appear when an ionizing group known as an *auxochrome* is attached. These ionizing groups may, of course, carry either a positive charge, as does the cation NH_2, or a negative charge, as does the anion SO_3. The ions in solution must, of course, be balanced, so that cationic dyes like methylene blue require the addition of a free anion—in this case chloride—while the anionic dyes are usually balanced by a sodium ion.

The term staining involves the concept that the dye must become attached to the structure to be stained and remain attached through the various manipulations inherent in the process of mounting. The commonest reason for the permanent attachment of the dye is that the charge on the dye becomes balanced by the charge on the particle dyed, so that basic dyes like methylene blue or hematoxylin become attached to those molecules which retain a negative charge after they have been fixed and sectioned. The obvious molecules in this instance are the nucleic acids, so that the two dyes mentioned are known as *nuclear stains*. Most of the proteins found in the cytoplasm, on the contrary, retain a positive charge, so that acid dyes like eosin remain attached to them. The terms basic and acidic in this connection have nothing whatever to do with the pH of the solution used in staining; they refer solely to the charge carried by the auxochrome ion. The extent to which a dye is acid or basic depends on the number of charged auxochromes and chromophores which enter into the composition of the molecule. Many dyes are insufficiently charged to remain attached through the necessary manipulations of microtechnique. The degree of adhesion can, however, be increased by an intermediary substance known as a *mordant*. Sulfates, particularly the double sulfates known as *alums*, are the most widely used mordants in microtechnique. In some cases it appears that the mordant becomes attached to the tissue by a covalent bond, and the dye is therefore rendered completely insoluble in neutral solutions used for dehydrating. Dyes attached to tissues by mordants may, however, be removed either in ionized solutions, usually acid, or with the aid of solutions of the mordant itself. This is the basis of the differentiation used with hematoxylin and carmine.

Another basis for differential staining is a simple difference in permeability between different types of tissue. Most blood stains are of this type, for the leucocytes are permeated easily by large molecules, while the erythrocytes are permeated only with difficulty even by small molecules. A mixture of a large blue molecule and a small yellow molecule is thus the basis of the standard blood techniques such as that of Wright.

Another less frequent method of differential staining depends on the solubility of various dyestuffs. For example, Sudan III is oil-soluble but water-insoluble, so that a colloidal dispersion of the dye in water is picked up by oil droplets in contrast to other tissues present. Staining, no matter what the theoretical basis of the individual process, may be either direct or indirect. In the former case, the dye is applied to the tissue or object, which is removed from it when sufficient dye has been absorbed. In indirect staining, the tissue or object is soaked in the dye solution and then differentiated, usually in an acid or mordant, until the dye has been removed from unwanted regions. Direct staining, particularly from very weak solutions, is usually preferable to indirect staining in the case of large wholemounts, while indirect staining is more usually employed on sections or small objects.

The first nuclear stains used by microscopists were hematoxylin, a reagent extracted from campeche wood, and carmine, a derivative of an extract of cochineal. These are still extensively used, even though there is today every reason for replacing such natural products with synthetic dyes. Hematoxylin is probably the commonest nuclear stain, and descriptions of mordant, direct, and indirect hematoxylin methods will therefore be given first.

HEMATOXYLIN

This dye is available as an artificially bleached, whitish powder obtained by solvent extraction from the residue of an aqueous extract of campeche wood. The dye is readily soluble in alcohol and water, but these solutions have little staining value except in mordant techniques. Staining solutions incorporating mordants require to be ripened for some months. The staining qualities of ripe solutions are in part due to the oxidation of hematoxylin to hematein. There must, however, be other factors involved, since neither the substitution of hematein for hematoxylin nor the addition of oxidizing agents produces a stain of the same quality.

Mordant Hematoxylin Staining. These techniques give a very intense nuclear stain, of more interest to nuclear cytologists than histologists. Ferric alum is the mordant most commonly used. Simple solutions of mordant and stain were used by Heidenhain, whose name is usually associated with the technique, but better results can nowadays be obtained by the technique of Régaud on regular sections or that of Dobell on very thin sections or protozoan smears.

Régaud's Mordant Hematoxylin Stain:

Mordanting solution
  5% iron alum
Staining solution
  Water   80 ml
  Hematoxylin   1 g
  Glycerin   10 ml
  Alcohol   10 ml
Differentiating solution
  Alcohol   65 ml
  Water   35 ml
  Picric acid   0.5 g

This solution is the best of the general-purpose iron hematoxylins and should be widely used to stain the nuclei before complex afterstaining. The staining solution must be ripened for some time before use. The most convenient method of doing this is to prepare a 10 per cent solution of hematoxylin in alcohol and then to dilute this with the glycerin in water immediately before use. The alcohol solution of hematoxylin should be at least a month old. This stain is very slow if it is used cold, so that it is customary to heat both the mordanting and staining solutions to about 50°C before use.

Sections Are Stained as Follows:

1. Accumulate the sections in distilled water.
2. Transfer the slides to the mordanting solution for 30 min at 50°C or overnight at room temperature.
3. Rinse each slide in distilled water to avoid carrying over too much of the mordant into the staining solution.
4. Transfer the slides to the staining solution for 30 min at 50°C or overnight at room temperature.
5. Transfer the slides to distilled water and wash until no more stain comes away.
6. Dip each slide up and down in the differentiating solution until it appears to be partly differentiated and transfer to tap water until no further color comes away. Then examine the slide under the microscope. If further differentiation is required, repeat the process.
7. Transfer all the slides to tap water until they have turned blue. If the tap water becomes yellow from traces of picric acid, it should be changed or differentiation will continue. Hematoxylin is very sensitive to acids and, as it comes from the differentiating solution, has a reddish-brown color, as distinct from the clear blue color required on the finished slide. In many parts of the world, natural tap water is sufficiently alkaline to produce this blue color. If it is not, a pinch of sodium bicarbonate should be added to the coplin jar containing the tap water.

Dobell's Mordant Hematoxylin Stain:

Mordanting solution
    1% iron alum in 70% alcohol
Staining solution
    1% hematein in 70% alcohol
Differentiating solution
    0.1% hydrochloric acid in 70% alcohol

All these solutions are stable, and it will be noticed that the staining solution uses hematein, so that no ripening is required. This method can be used only on very thin sections or on protozoan smears because it gives such an intense coloration that thick sections can be differentiated only with difficulty.

Procedure Is as Follows:

1. Accumulate thin sections or protozoan smears in 70 per cent alcohol.
2. Transfer them to the mordanting solution for about 10 min.
3. Without rinsing, transfer the sections directly to the stain for 10 min.
4. Rinse them quickly in 70 per cent alcohol.
5. Differentiate sections in the last solution until only the chromosomes or nuclei remain stained. Differentiation is fairly rapid. Remove sections from the differentiating solution at intervals, wash thoroughly in 70 per cent alcohol, and then examine under the microscope. If a number of slides are being stained at the same time, it is desirable to differentiate one carefully, noting the time needed for the operation, and to use this same amount of time to bring the remaining sections through in one batch.
6. Wash them thoroughly in 70 per cent alcohol.
7. Counterstain the sections, if desired, by one of the methods to be described later or mount them in balsam after dehydration and clearing.

Direct Hematoxylin Staining. The solutions used for these techniques contain both dye and mordant in one solution. The commonest mordants are either potassium or ammonium alum, and most formulas contain glycerin as a stabilizing agent. Literally hundreds of mixtures have been proposed, but they may be broadly divided into acidified and nonacidified mixtures. The former are generally used in histological and pathological routines to stain nuclei before eosin counterstaining; the latter are used, although less frequently, for the same purpose and for staining xylem in plant sections.

Both types of stain must be blued after differentiation. The acid solution used for differentiation leaves the nuclei colored red, in which form the dye is unstable. Exposure to alkali changes the dye to a dark-blue stable form.

Ehrlich's Acid Alum Hematoxylin:

Staining solution
Water	30 ml
96% alcohol	30 ml
Glycerin	30 ml
Glacial acetic acid	3 ml
Hematoxylin	0.7 g
Ammonium alum	to excess

Note: Make up the stain by first dissolving the hematoxylin in the acid and alcohol. Then dissolve 1 g of ammonium alum in the water and add this together with the glycerin. Shake well and add about 10 g of ammonium alum to the bottle. Allow to ripen some months, making sure that there are always a few crystals of ammonium alum in the bottom of the bottle. This stain keeps well for about 10 years and works best after it is a year old. Large laboratories, therefore, should make up half a gallon every 2 or 3 years in order to have well-ripened material always in stock.

Bluing solution
    0.1% sodium bicarbonate in tap water

Stain Is Used as Follows:

1. Accumulate the sections in 90 per cent alcohol.
2. Place each slide for ½ to 2 min in the staining solution.
3. Remove each slide and drop 90 per cent alcohol on it from either a drop bottle or a pipette until the excess stain has been removed.
4. Transfer the slides directly to alkaline tap water until blue.

Note: If the sections are taken directly from water to stain or from stain to water, a diffuse stain that is very difficult to differentiate will result.

Delafield's Alum Hematoxylin:

Staining solution
Water	70 ml
Glycerin	15 ml
Methanol	15 ml
95% alcohol	4 ml
Ammonium alum	3 g
Hematoxylin	0.6 g

Note: Dissolve the dye in the alcohols and the alum in the water. Mix the two solutions and ripen for some months.

Differentiating solution
  0.1% hydrochloric acid in 70% alcohol
Bluing solution
  0.1% sodium bicarbonate in tap water

Method for Sections

1. Accumulate slides in tap water.
2. Stain until sections are dark purple. About 5 min for animal tissues or several hours for plant sections.
3. Rinse in distilled water.
4. Differentiate until only nuclei or xylem remains colored.
5. Blue until nuclei or xylem is dark blue, usually from 1 to 2 min.

Indirect Hematoxylin Staining. Indirect staining is that method in which very dilute stain is applied from a solution containing a mordant. The stain builds up slowly on the nuclei but is not absorbed by the cytoplasm. This method is never used on sections but is excellent for wholemounts. Any alum hematoxylin may be used in the following method:

Staining solution
  Delafield's hematoxylin   1 ml
  1% ammonium alum   20 to 100 ml

Note: The larger the object, the more dilute must be the stain. A protozoan can be adequately stained in a 1:20 dilution. A liver fluke would require at least a 1:100 dilution.

Washing solution
  1% ammonium alum
Bluing solution
  0.1% sodium bicarbonate

Method for Wholemounts

1. Accumulate fixed and well-washed objects in distilled water.
2. Place in stain until sufficiently stained, that is, until internal structures are clearly seen.
3. Wash in washing solution until no more color comes away.
4. Blue in bluing solution until all dye is converted to blue form. This requires at least 5 min for a small hydroid and as much as 12 hr for a large liver fluke. Lack of adequate bluing is responsible for most failures with this technique.

CARMINE

This substance is a compound of carminic acid, derived from the cochineal insect, with tin and aluminum compounds. The dye is not soluble in water or alcohol but is readily soluble in strong alkalis (e.g., ammonium hydroxide) and slightly soluble in weak alkalis (e.g., borax) or weak acids (e.g., acetic or propionic). Carmine staining solutions are no longer used for sections but are widely employed as indirect stains for wholemounts or as direct stains for squash preparations.

Indirect Carmine Staining. Both of the carmine formulas given below can be used as direct stains if they are diluted several hundred times with an alum solution. It is not safe to dilute them with distilled water, for there is a risk of the carmine precipitating out, particularly within the cavities of whole animals, from which it is almost impossible to remove it subsequently. The best known of all the carmine nuclear stains is:

Grenacher's Alcoholic Borax Carmine:

Staining solution
Boil together 10 g of borax and 8 g of carmine in 250 ml of water for about 30 min. Cool overnight and filter. Evaporate the filtrate to dryness and then store this powder, labeled as "Borax Carmine Powder." This powder may be dissolved in any strength alcohol from 30 to 70%; in all cases a saturated solution should be employed. This is the original method of making up the solution and is undoubtedly the best.

For those, however, who do not wish to take the time for the later preparation, a working solution may be made directly in the following manner:

Boil 2 g of borax and 1.5 g of carmine with 50 ml of water for 30 min. Cool and then add 50 ml of 70% alcohol. This solution should be filtered after 2 or 3 days.

Method of Using the Solution Prepared from the Dry Stock Powder Is as Follows:

1. Accumulate the objects in whatever percentage of alcohol is convenient.
2. Make up a saturated solution of the dry powder in alcohol from 10 to 20 per cent stronger than that in which the objects are accumulated.
3. Transfer the objects to stain until they have become a deep red color. This will take from 5 min for an individual protozoan to overnight for a medium-sized flatworm.

4. Transfer the objects to 0.1 per cent hydrochloric acid in alcohol of the same strength as that used for making up the stain. Let them remain in this solution until they become pink and translucent.
5. Dehydrate and mount the objects in the usual manner. This stain is not only good, used in the manner described, for staining wholemounts but is also much the best method of staining the nuclei in blocks of tissue before sectioning if this technique is to be employed.

Grenacher's alcoholic borax carmine is not very satisfactory when diluted and used by the indirect method. For this purpose it is recommended that the following be employed:

Mayer's Carmalum:

Staining solution
Boil together, for 1 hr, 2 g of carmine and 5 g of potassium alum in 100 ml of water. Cool and filter.

This solution may be used exactly as Grenacher's alcoholic borax carmine for wholemounts or blocks of tissue, but it is less satisfactory for this purpose. It is best used after great dilution for direct staining of relatively large invertebrates.

Method to Be Employed Is as Follows:

1. Accumulate the objects in distilled water.
2. Dilute the stain with 5 per cent potassium alum to the required concentration. The larger the object to be stained, the lower the concentration of the staining solution should be. For a relatively small and thin object, such as the prothallium of a fern, a dilution of about 10 to 1 is satisfactory. For a large object, such as a liver fluke, the stain should be diluted until it is only faintly pink.
3. Transfer the object to the stain and leave it until examination shows the nuclei or internal structures to have been stained a fairly dark red, while the other portions are stained only pink. This will take about 6 hr with a fern prothallium in the dilution described or about 3 weeks for a liver fluke at the low concentration.
4. Wash the object in running water until all alum has been removed and then mount in the ordinary manner.

ORCEIN

Many lichens contain a crude pigment which, even as late as the early nineteenth century, was used to dye textiles. An oxidation product of one of the ingredients of this pigment is used by cytologists, under the name orcein, to stain chromosomes in squash preparations. It has no other use. The following technique is the best known:

LaCour's Aceto-Orcein:

Water	20 to 50 ml
Acetic acid	70 to 30 ml
Orcein	1 g

Note: The orcein is dissolved in the dilute acid, the concentration of which is varied according to the material to be stained. Fifty per cent acetic acid is commonly used by plant cytologists; 70 per cent acetic acid is preferred for Drosophila.

Method for Squashes

1. Place freshly dissected living material in a drop of stain.
2. Leave 3 to 10 min.
3. Apply coverslip and squash to shatter cells and display chromosomes.

SYNTHETIC NUCLEAR STAINS

Of the four stains given below, only celestine blue B and safranin are usually considered to be true nuclear stains, the other stains having been developed for staining bacteria. However, they are all considered together since the bacterial stains often can be used to demonstrate nuclei in material in which conventional stains break down. For example, when one is staining a section of a frog larva or egg, which is heavily loaded with yolk, one has great difficulty in endeavoring to use hematoxylin for the reason that some of the albuminous yolk particles pick up this stain. Either carbolmagenta or crystal violet, however, will demonstrate the nuclei clearly without being picked up by the yolk.

Celestine Blue B. This is much the most satisfactory synthetic nuclear stain for sections since a stable staining solution is now available. Early staining solutions were both unstable and erratic in their action and have given this dye a bad name, but the following method is foolproof. It is the only direct nuclear stain available for sections.

Gray's Celestine Blue B:

Dissolve 2.5 g iron alum in 100 ml water with 14 ml glycerin. Place 1.0 g celestine blue B in a beaker. Tilt the beaker and tap to accumulate the dye in one place. Pour on 0.5 ml concentrated sulfuric acid and mix with a glass rod. When effervescence has ceased, the dye will be in the form of a friable mass. Break up this mass coarsely and pour on, with constant stirring, the iron alum solution heated to 50°C. Cool to room temperature *and adjust to pH 0.8 with concentrated sulfuric acid.*

Note: Both the method of compounding and the pH are critical. At pH 0.6 there is no staining. At pH 1.0, nuclei are well stained but the cytoplasm picks up some stain. At pH 1.2 the cytoplasm is rather densely stained. Those who lack a pH meter and a magnetic stirrer, which is the only really satisfactory combination for adjusting the pH, should add 1.5 ml concentrated sulfuric acid to the cooled solution. After thorough mixing, stain a test slide. If the cytoplasm is colored, mix in a few more drops of acid and test again; continue until the critical point is reached. Usually about 1.8 ml of acid is required.

Method for Sections

1. Accumulate slides in water.
2. Place in stain from 1 to 5 min. The time is not critical since overstaining cannot occur.
3. Rinse in water, dehydrate, clear, and mount in balsam.

Safranin. Safranin nuclear staining in English-speaking countries is usually confined to botanical specimens, though its use in European countries is widespread for histological purposes. Safranin is not easy to use and is a very slow stain. Undoubtedly the best method is:

Johansen's Safranin:

Staining solution
Methyl Cellosolve	50 ml
95% alcohol	25 ml
Water	25 ml
Sodium acetate	1 g
40% formaldehyde	2 ml
Safranin	0.1 g

Note: Dissolve the dye in the methyl Cellosolve. Then add the alcohol. Dissolve the sodium acetate in the formaldehyde and water and add these to the dye solution.

Differentiating solution
Identical with that used with Régaud's hematoxylin (see page 100).

Method for Plant Sections

1. Accumulate slides in 70 per cent alcohol.
2. Stain for 1 to 3 days.
3. Differentiate until nuclei and xylem are sharply defined.

Stain Is Used on Animal Tissues as Follows:

1. Accumulate the sections in water.
2. Transfer the sections to stain and let them remain there from 24 to 48 hr (or even longer) until the nuclei are darkly stained.
3. Dip the slide up and down in the differentiating solution until the unwanted stain has been removed from the cytoplasm.
4. Wash it in running water to remove the picric acid.
5. Mount the section in the usual manner.

The stain given below is the basis for about 90 per cent of all bacteriological techniques.

Other Synthetic Nuclear Dyes

Ziehl's Carbolmagenta:

Staining solution
Grind together 1 g of magenta (basic fuchsin) with 10 ml of 90% alcohol and 10 g of phenol. When this mixture has been reduced to a paste, take 100 ml of water and rinse out the mortar, using 10 successive 10-ml batches. These batches should be accumulated, left a few hours, and then filtered.

The use of this stain for bacteria is described on page 227. If it is to be used for staining nuclei in either animal or plant sections, it is necessary only to transfer the section from water to the stain for 10 to 20 min and then to differentiate as long as is required with 1 per cent acetic acid.

It is sometimes necessary to stain bacteria blue, either as a contrast to the red color produced by Ziehl's solution or in order to differentiate bacteria that have been decolorized. Much the best solution for this purpose is:

Lillie's Ammonium Oxalate Crystal Violet:

Note: This solution is often called *Hucker's ammonium oxalate crystal violet* or *Hucker-Conn crystal violet.*

Staining solution
Dissolve 2 g of crystal violet in 20 ml of 95% alcohol. Dissolve separately 0.8 g of ammonium oxalate in 80 ml of distilled water. When both solutions are complete, add the oxalate solution to the dye solution.

The proportions used here are those given by Lillie.

The use of this solution in staining bacterial films is described on page 222, and its use for the staining of nuclei is identical with the method outlined for Ziehl above.

PLASMA OR CONTRAST STAINS

Wholemounts are almost invariably stained in one color only since sufficient nuclear stain usually remains dispersed throughout the plasma to provide adequate visibility. The cytoplasm of sections, however, is usually stained a contrasting color, both to render the nuclei more apparent and to emphasize the general structure. Sometimes two or more colors may be employed. Single contrasts are usually perfectly adequate, and the following may be recommended:

Single-contrast Stains

Eosin Y:

Make up as a 0.5% solution in distilled water.

Ethyl Eosin:

Make up as a 0.5% solution in 95% alcohol.

Both these eosins give very much the same shade and are the conventional contrast to hematoxylin- or celestine blue B–stained sections. The choice between water and alcohol is a matter of individual preference.

Another popular contrast stain is:

Fast Green FCF:

Make up as a 0.1% solution in 90% alcohol.

This is sometimes used as a contrast for hematoxylin, for which purpose it is not so satisfactory as the eosins, but it is very widely employed as a contrast stain to red nuclei stained with either safranin or carmine.

Double-contrast Stains. It is no more trouble to use a single solution that will stain different tissues various shades than it is to use a simple solution. Two solutions can be confidently recommended. The first of these is:

van Gieson's Stain:

Staining solution
To 100 ml of a saturated solution of picric acid in water, add 0.05 g of acid fuchsin.

Stain Is Used as Follows:

1. Collect sections, with the nuclei stained blue, in tap water. Celestine blue B is the stain of choice.
2. Stain them from 2 to 5 min in the staining solution.
3. Give each slide an individual quick rinse in tap water and pass it directly to 96 per cent alcohol.
4. Keep the slides in motion in 96 per cent alcohol until they are dehydrated.
5. Then rinse each section once or twice in absolute alcohol and transfer it to xylene.
6. As soon as the section is clear, mount it in balsam.

This stain relies for its effect on the fact that muscular tissues retain the yellow color of the picric acid more readily than the other connective tissues. The major objection is that picric acid is soluble to a certain extent in xylene so that the yellow color is gradually extracted. Sections stained by this method and mounted in balsam are rarely of very much use a year after they have been prepared. To avoid these objections, the author much prefers:

Gray's Double-contrast Stain:

Staining solution
  Water   100 ml
  Orange II   0.6 g
  Ponceau 2R   0.4 g

Solution Is Used as Follows:

1. Accumulate sections, with nuclei stained either in hematoxylin or in celestine blue B, in tap water.
2. Transfer the slides to the staining solution for 2 min.
3. Remove each slide individually from the stain, drain, blot, and then dip up and down in absolute alcohol until sufficiently differentiated. The completion of differentiation and the completion of dehydration usually coincide.
4. Then transfer each slide to xylene and mount in the ordinary manner.

This stain gives a good range of red-orange and gold shades on most histological sections and is no more difficult to use than a simple solution of eosin.

Double-contrast stains for sections in which the nuclei have been stained red are not very common, though the following, originally designed for use with heavily yolked embryonic material, is really excellent:

Smith's Picro–Spirit Blue:

Staining solution
    Absolute alcohol   100 ml
    Picric acid   1 g
    Spirit blue   *enough to saturate*

Stain Is Used as Follows:

1. Take sections of material which have been bulk stained in carmine through the ordinary procedures as far as absolute alcohol.
2. Transfer them to the staining solution for 2 min and rinse individually in absolute alcohol until sufficiently differentiated.
3. Then transfer the slides to xylene, which stops further differentiation. This gives good differentiation of a large number of tissues. On embryonic material it is particularly effective, for the nuclei are red and yolk and yolk granules are yellow-green to green, while yolk-free cytoplasm is a clear blue.

Complex Staining Techniques for Animal Tissues. Complex staining solutions are those in which a series of stains, mordants, and differentiating solutions is used one after another in such a manner that the nuclei and all the elements of the plasma are stained in sharply contrasting colors. The stains used in botany, naturally, are quite different from those used in histology and pathology and are given immediately following this section. Many hundred complex stains are in the literature and will be found in Gray's "Microtomist's Formulary and Guide," where several hundred more stains of specific application are also given. Here it is intended to present only three stains, which are so simple to use and so excellent in their results that they should be known to everyone. The first of these is:

Mallory's Triple Stain:

First staining solution
    1% acid fuchsin
Differentiating and mordanting solution
    1% phosphotungstic acid
Second staining solution
    Water   100 ml
    Aniline blue   0.5 g
    Orange G   2 g
    Oxalic acid   2 g

Method of Use

1. Accumulate sections for staining in water.
2. Stain them in the first staining solution for 2 min.
3. Rinse the slides thoroughly in water.
4. Transfer them to the phosphotungstic acid for 2 min.
5. Give the slides a very quick rinse in water. The purpose is to remove the phosphotungstic acid from the slides, not from the sections.
6. Transfer to the second staining solution for 15 min.
7. Wash the slides in water until no more color comes away.
8. Take each slide individually and dip it up and down in absolute alcohol until it is differentiated. This may be seen clearly with the naked eye, for the slide, which is a muddy purple when differentiation starts, suddenly clears to show bright-blue and bright-red areas.
9. Transfer the slide to xylene, which stops differentiation.

When using this stain for the first time, it is well to transfer sections to xylene and examine them under the microscope before differentiation is complete and then to put them back into absolute alcohol to complete the differentiation. A successful slide will show nuclei in red, cartilage and white fibrous connective tissue in blue, nerves and glands in various shades of violet, muscle in red, and erythrocytes and keratin in orange. The only disadvantage of the technique is the rapidity with which the blue color is removed in absolute alcohol, so that differentiation must be watched very carefully. The dyes used are extremely sensitive to alkali, so that the slides, if they are to be kept for some time, should be mounted in a very acid medium. A procedure that has been recommended is to keep a saturated solution of salicylic acid in xylene and to dip each coverslip in this before applying it to the finished preparation. This is much less trouble than, and just as effective as, making up a special salicylic acid balsam.

Another complex contrast stain, in which the nuclei first must be stained with hematoxylin, is:

Patay's Triple Stain:

First staining solution
    1% Ponceau 2R
Differentiating and mordanting solution
    1% phosphomolybdic acid
Second staining solution
    0.5% light green in 90% alcohol

Method of Use

1. Stain the sections in Delafield's hematoxylin (see page 102) and differentiate until not only the nuclei but also the cartilage remains blue. Wash the sections in

alkaline tap water until they are blue and then wash thoroughly in distilled water to get rid of all the alkali.
2. Put the sections in the first stain for 2 min.
3. Rinse them briefly in water.
4. Place them in the differentiating and mordanting solution for 2 min.
5. Rinse sections briefly in water.
6. Transfer them to 95 per cent alcohol until dehydrated.
7. Transfer each section individually to the second staining solution for 30 sec.
8. Rinse each slide individually in absolute alcohol until no more color comes away.
9. Transfer the slides to balsam and mount in the usual manner.

This is one of the most brilliant of all the triple stains, and a successful preparation shows the nuclei in blue-black and the cartilage in clear blue. Other connective tissues are green, white fibrous connective tissue being light green, and bone is a most brilliant green in contrast to the orange muscle and blue cartilage. Red blood cells are yellow, and most nerve tissues are colored a neutral gray. This stain is little more trouble to apply than Mallory's, but the results are both more brilliant and more permanent.

Many complex stains rely on the fact that a mixture of eosin and methylene blue solutions gives rise to a precipitate which, though insoluble in water, is soluble in methyl alcohol. These stains are used mainly for blood films, not only to differentiate the cell types but also to display the parasites present. The best known of these is:

Wright's Stain:

The preparation of this stain is very complicated. (It is described in Gray's "Microtomist's Formulary and Guide.") It is strongly recommended that the solution be purchased and used as it comes from the bottle.

Method, as Applied to Air-dried Blood Smears, Is as Follows:

1. Flood the stain on the slide from a drop bottle and leave for 2 min.
2. From another drop bottle, add distilled water drop by drop until a green scum forms on the surface of the stain.
3. Wash slide in distilled water until no further color comes away.
4. Dry and examine the slide.

This is the standard stain used for counting and differentiating leukocytes. There are hundreds of variations to this method, and also hundreds of other methods of staining blood. Wright's stain is, however, the basic method from which most others have been derived.

The use of Wright's stain is almost invariably confined to blood smears. Methylene blue–eosin techniques may be applied to sections, usually for the purpose of displaying bacteria in pathological material. The method of Mallory is standard:

Mallory's Methylene Blue–Azure II–Phloxine:

First staining solution
  2.5% phloxine
Stock second staining solutions
  A. Methylene blue   1
     Sodium diborate 1
     Water   100
  B. 1% azure II
Working second staining solution
    Stock A   5
    Stock B   5
    Water   90
Differentiating solution
    95% alcohol   100
    Resin   0.5

Method of Use

1. Accumulate slides in distilled water.
2. Place slides in a coplin jar of first staining solution. Place in oven at 55°C for 1 hr. Remove from oven and cool to room temperature.
3. Rinse *briefly* in distilled water.
4. Place in second staining solution for from 5 to 20 min or until sections assume bluish tinge.
5. Differentiate in differentiating solution until nuclei are dark blue on pink background.
6. Rinse in 95 per cent alcohol.
7. Dehydrate in absolute alcohol and clear in two changes of xylol.

Complex Botanical Stains. Most of the combinations used for zoological staining, such as hematoxylin-eosin, can also be used in plant structures, where it is desired to differentiate between the nucleus and the cytoplasm. It is only a matter of convention that the safranin and light green combination, so little used in zoological techniques, is preferred by most botanists for their material.

The complex stains are an altogether different matter, since the chemical nature of plant structures is naturally different from that of animals. The best of these complex botanical stains is:

Johansen's Quadruple Stain:

First staining solution
  Johansen's safranin

Second staining solution
  1% methyl violet 2B
First differentiating solution
  95% alcohol   30 ml
  Methyl Cellosolve   30 ml
  Tertiary butyl alcohol 30 ml
Third staining solution
  Mix 6 ml of methyl Cellosolve with 6 ml clove oil and saturate this mixture with fast green FCF. Filter the saturated solution and add to it 35 ml of 95% alcohol, 35 ml of tertiary butyl alcohol, and 12 ml of 1 per cent acetic acid.
Second differentiating solution
  95% alcohol   50 ml
  Tertiary butyl alcohol 50 ml
  Glacial acetic acid   0.5 ml
Fourth staining solution
  Prepare separately saturated solutions of orange G in methyl Cellosolve and 95% alcohol. Mix 50 ml of each solution.
Third differentiating solution
  Clove oil   30 ml
  Methyl Cellosolve   30 ml
  95% alcohol   30 ml
Special dehydrating solution
  Clove oil   30 ml
  Absolute alcohol   30 ml
  Xylene   30 ml

Stain Is Used as Follows:

1. Accumulate sections mounted on slides in 70 per cent alcohol and place them in the first staining solution for 1 to 3 days.
2. Wash slides in running tap water until no more color comes away.
3. Transfer them to the second staining solution for 10 to 15 min.
4. Rinse slides in running tap water.
5. Differentiate them for 10 to 15 sec in the first differentiating solution.
6. Stain sections in the third staining solution from 10 to 20 min or until sufficient green dye has been absorbed.
7. Rinse them for 5 to 10 sec in the second differentiating solution.
8. Place sections in the fourth staining solution from 3 to 5 min or until the cytoplasm of the cells has become bright orange.
9. Dip them up and down three or four times in the third differentiating solution.
10. Rinse for 10 to 15 sec in special dehydrating solution.
11. Transfer slides to xylene and mount in the ordinary manner.

In a successful preparation of this type, chromosomes and lignified cell walls are stained bright red, the contents of the cells having purplish resting nuclei against a bright-orange cytoplasm. There is a very vivid, but somewhat variable, differentiation of the remaining structures. Parasitic fungi are particularly well shown in bright green whether they are penetrating cytoplasm or the lignified portion of the cell wall.

STAINS FOR SPECIAL PURPOSES

The number of stains for special purposes is legion. They belong properly in the field of the specialist in the particular tissue that they demonstrate. There are, however, a few that are sufficiently interesting and simple to justify their inclusion in a beginner's handbook.

Fats. Most fats are normally dissolved from tissues in the course of embedding in wax. If, however, frozen sections are made by the technique described on page 185, it is possible to demonstrate differentially the fat globules by soaking the sections in an alcoholic solution of a fat-soluble, but water-insoluble, dye. The classic method is to use a saturated solution of Sudan IV in 70 per cent alcohol. This stains fat globules red. A blue color may be obtained by using a saturated solution of oil blue N in 60 per cent isopropyl alcohol. Sections stained by this method cannot be dehydrated and should be mounted in Farrants's medium.

Skeletons. It is often useful, in the study of embryos or very small vertebrates, to be able to stain the skeleton differentially.

Specimens, such as fish fry, which have bony skeletons, should be preserved in 70 per cent alcohol made slightly alkaline by saturating it with borax. When they are thoroughly hardened, a 0.5 per cent solution of alizarine red S in absolute alcohol is added in the proportion of 1 ml of stain for each 100 ml of preservative. The alizarine forms a red lake with the calcium in the bones. When the bones are red enough, the surplus stain is washed out of the other tissues with alkaline alcohol. The embryos are then dehydrated and mounted in balsam.

Cartilaginous skeletons cannot be directly stained, as are bones, but may be indirectly stained in toluidine blue. The embryos are fixed in any fixative not containing picric acid—mercuric mixtures are preferred by most people—and thoroughly washed. They are soaked for 24 hr in:

Van Wijhe's Stain:

70% alcohol	100 ml
Hydrochloric acid	0.1 ml
Toluidine blue	0.1 g

Then they are differentiated in 0.1 per cent hydrochloric acid in 70 per cent alcohol until no more color comes away. Dehydration and mounting in balsam will show the cartilage alone stained clear blue.

SUGGESTED ADDITIONAL READING

Baker, J. R.: "Principles of Biological Microtechnique," New York, John Wiley & Sons, Inc., 1958.
Conn, H. J.: "Biological Stains," 7th ed., Baltimore, The Williams & Wilkins Company, 1961.
Gray, P.: "The Microtomist's Formulary and Guide," New York, McGraw-Hill Book Company, 1954.
Gurr, E.: "Encyclopaedia of Microscopic Stains," Baltimore, The Williams & Wilkins Company, 1960.
——: "Staining—Practical and Theoretical," Baltimore, The Williams & Wilkins Company, 1962.

CHAPTER 8 | Dehydrating and Clearing

General Principles. Dehydration means "the removal of water from." It is a necessary step in the preparation of specimens for microscopic examination because most of the media in which they will be mounted are not miscible with water. The term clearing is rather misleading. It is used in microscopy to describe the process of removing alcohol from dehydrated tissues. This process is necessary because the resinous media used for mounting many specimens and waxes used for embedding before cutting sections are no more miscible with alcohol than they are with water. Many of the reagents used for clearing have a high index of refraction so that they make objects saturated with them appear more transparent or clearer. Most of the reagents used for dehydrating and clearing cause considerable shrinkage of tissues. This does not matter in the case of animal tissues since the shrinkage is uniform and does not much distort the appearance of the specimen. Shrinkage in plant tissues must, however, be avoided at all costs because the cellulose walls do not contract as much as the cell contents. It follows that a carelessly dehydrated or cleared plant tissue will show the contents of the cell pulled away from the wall.

Ethyl Alcohol. The commonest reagent used for dehydrating is ethyl alcohol, which is available in most laboratories as both neutral grain spirits (95 per cent alcohol) and absolute alcohol (100 per cent alcohol). The removal of the last 5 per cent of water from neutral grain spirits is a very expensive operation, so that 95 per cent alcohol should be used wherever possible.

Were the majority of specimens merely to be thrown in 95 per cent alcohol, the violent diffusion currents that would be set up would result in the collapse of cavities or in the distortion of the specimen. It is customary, therefore, to use these alcohols as a graded series. It is conventional today to employ the series of 30 per cent, 50 per cent, 70 per cent, 90 per cent, and 95 per cent and to pass the specimen from one of these strengths to the next, leaving it in each sufficiently long to become impregnated. This series is not reasonable, for there is a much greater and more violent diffusion current when a specimen is passed from water to 30 per cent alcohol than there is when a specimen is passed from 70 per cent to 90 per cent alcohol. The author much prefers to use the series 15 per cent alcohol, 40 per cent alcohol, 75 per cent alcohol, and 95 per cent alcohol and would recommend this for the beginning student. This series more nearly represents the intention of the worker, which is to subject the specimen to a graded series of stresses rather than to a graded series of alcohols.

It is doubtful whether or not it is necessary even to use a series of alcohols when the object to be dehydrated is a thin section attached to a slide. The only purpose of using an intermediate concentration of alcohol between water and 95 per cent is to avoid the rapid dilution of the latter by the water carried over on the surface of the slides. This difficulty can be avoided by using two jars of 95 per cent alcohol. It must be understood, of course, that the section will take just as long to dehydrate as when a series is employed, but one avoids the difficulty of transferring the section through many jars.

Other Dehydrants. Many substitutes for ethyl alcohol have been proposed, some of them intended for use in circumstances where ethyl alcohol is hard to obtain and others intended to serve the purpose of a universal solvent, miscible alike with water, balsam, and wax. The best of the alcohol substitutes is Cellosolve (ethylene glycol monoethyl ether). This substitute has many advantages over ethyl alcohol for purposes of dehydration but cannot be substituted for it in the preparation of many stains and staining solutions. It is less volatile than alcohol, so that, if left in an uncovered dish, it does not evaporate so rapidly. It is also somewhat less hygroscopic, so that, under the same circumstances, it does not lose its strength. It has the disadvantage that it is more viscous than alcohol and tends to give rise to greater diffusion stresses; therefore, it must be used in a more extended series of graded mixtures if delicate wholemounts are to be passed through it. However, for the handling of tissues intended for embedding in paraffin and, above all, for the routine handling of tissues in pathological laboratories, Cellosolve is strongly recommended.

The only one of the universal solvents that has found any general acceptance is dioxane (diethylene dioxide). This solvent is readily miscible with water and with balsam and is slightly less miscible with molten paraffin. Specimens, therefore, can be transferred directly from water to dioxane—a graded series is necessary should they be delicate—and, after having been thoroughly impregnated with dioxane, transferred directly either to the mounting medium or to a bath of molten paraffin for impregnation. In spite of the apparent simplicity of the use of this solvent, there are certain great objections. The first of these is the toxicity of dioxane vapor to humans. Dioxane is a cumulative poison and has been shown to affect seriously the function of both the liver and the kidneys. This does not matter very much where it is used by an individual who knows himself to be free from hepatic or renal disorder, particularly where he is exposed only to low concentrations of the vapor for relatively short periods. However, it militates heavily against the use of this reagent in large classes, where the instructor is responsible for the health of individuals without knowing their physical idiosyncrasies. Another disadvantage is in the heavy diffusion stresses that are set up when materials are transferred from dioxane to molten paraffin. This is not so important when dealing with tissue blocks for routine histological examination, but it is almost impossible to get a good section of a 72-hr chick embryo, for example, using this shortened dioxane technique.

As has already been pointed out, dehydrating and clearing plant tissues are altogether different processes. Tissues may be run through ethyl alcohol, using a very long series of gradually increasing strengths, but it is more usual nowadays to employ mixtures of alcohol and tertiary butyl alcohol. Tertiary butyl alcohol is miscible with molten paraffin, so that dehydration and clearing take place at the same time in these graded series. This is described in more detail in Chapter 12, which discusses the process of cutting sections.

Clearing Agents. Clearing agents, which remove the dehydrating agent from the tissues and leave them in condition for either mounting in balsam or embedding in paraffin, are of two main types. For mounting in balsam, it is customary to use one of the essential oils or their synthetic equivalents. The advantage of these materials is that many of them are readily miscible with 90 to 95 per cent alcohol and, therefore, are capable of removing the last traces of water that may be left in the specimen after imperfect dehydration.

In the author's opinion, the best clearing agent for general use before making wholemounts is terpineol (synthetic oil of lilac). This material is readily miscible with 90 per cent alcohol and has the additional advantages that it neither has an unpleasant odor nor does it render objects brittle. It is, however, more customary today to recommend clove oil. This has the advantage of being much more fluid than terpineol and the disadvantages of a very pungent odor and the tendency to make small objects brittle. The latter is sometimes an advantage, as when one is endeavoring to remove appendages from small arthropods, but on other occasions it is very annoying. Many other oils have been recommended from time to time, but these two between them will be satisfactory for making a wide variety of wholemounts. Under no circumstances whatever should an essential oil be used to clear objects intended for embedding in paraffin because the oil is almost impossible to remove completely and will destroy the good cutting qualities of the embedding medium selected.

For clearing or dealcoholizing objects intended for embedding in wax, it is customary to use a hydrocarbon, and xylene is in almost universal employment at the present time. From the point of view of physical properties, there is little choice between benzene, xylene, and toluene; the author considers that the first of these has a distinct advantage in that it tends to render objects less brittle. All three are equally miscible with molten paraffin, but none is so good a solvent of solid paraffin as chloroform. Both the hydrocarbons and chloroform are very sensitive to water, so that it is essential that an object be completely dehydrated in alcohol or dioxane before being transferred to one of these clearing agents. It occasionally happens that it is impossible to provide perfect dehydration, and then one is forced to utilize the coupling properties of phenol, which is usually employed in the following mixture:

Carbolxylene:

Xylene   70 ml
Phenol   30 g

The exact proportion of the ingredients varies somewhat according to preference, but the proportion given above is the most customary. Specimens may be taken directly from 50 per cent alcohol into this reagent, which, however, must be removed thoroughly by washing in pure xylene before the specimen is transferred to wax. It should be pointed out to the inexperienced student that phenol is capable of giving a most unpleasant burn and that carbolxylene itself should be kept away from the hands under all circumstances. This reagent is also useful when one is endeavoring to clear objects in hot humid weather in summer. The alcohol used for the dehydration absorbs water so rapidly under these circumstances that it is almost impossible to clear satisfactorily in xylene alone.

In the case of small objects, no special precautions need to be observed in the technique of dehydrating. That is, one may merely place the object at the bottom of the tube, fill the tube with the required strength of alcohol, and change this for stronger alcohol as often as becomes necessary. This technique, however, cannot be employed if one is dealing with large objects, since there is a tendency for the water abstracted from the object to accumulate at the bottom of the tube and thus prevent satisfactory dehydration. Objects of a size larger than a grain of corn should always be suspended from the top of the vessel containing the dehydrating agent either in a small bag of cloth or, where the nature of the object permits, from a hook inserted into the cap of the tube or jar. These remarks do not apply when one is dehydrating with Cellosolve, which is denser than water.

The technique of clearing specimens in a hydrocarbon or in chloroform before embedding in wax is exactly the same as dehydrating, and no special precautions need to be observed. In the case of delicate objects intended for wholemounting, however, it is necessary to provide some gradient between the alcohol and the essential oil if one is to avoid distortion. The simplest way of doing this is to use a flotation method; that is, one first of all pours a layer of the oil on the bottom of the tube and then very carefully, with the utmost precautions against mixing, floats a layer of alcohol on top of this. The object is transferred from the alcohol to the upper layer of alcohol in the tube, through which it drops to the junction of the two fluids. As soon, however, as it has become partially impregnated with the clearing agent, it sinks to the bottom of the tube. As it lies there, columns of alcohol will be seen rising from it. When the alcohol has ceased to rise, the object is extracted with a pipette with care so as not to get any alcohol into the pipette and then transferred to a tube or dish of the pure essential oil, where clearing will be completed in a few minutes.

There is no more common cause of failure in the preparation of microscope slides than imperfect dehydration and clearing. It is a sheer waste of time to endeavor to embed an object in paraffin unless all the water and all the alcohol have been removed from it, and it is a waste of time to endeavor to impregnate it with xylene unless all the water has been removed from it by alcohol. It is quite impossible to give any particular time schedule for any particular object; experience is the only guide. It is easy to see, however, when an object has not been dehydrated or cleared perfectly. The least trace of milkiness—as distinct from opalescence—is an indication that the water has been imperfectly removed in alcohol, so that the object cannot be cleared properly. If this slight milkiness is observed, the object must be returned to absolute alcohol until such time as all the water has been removed. There is no simple method of determining when all the alcohol has been removed by xylene. It is usually safer to use three changes, allowing ample time in each, than to embed an object that has been cleared imperfectly and that will be impossible to section subsequently.

Perfect clearing is just as essential in objects intended for wholemounts and is much easier to determine, since the essential oils are of sufficiently high refractive index to make a properly cleared object appear glass-clear. No further clearing will take place in Canada balsam, so that, unless the object appears perfect in oil, it is a waste of time to mount it.

A final point to remember is that dehydrating agents must of necessity be hygroscopic and that they will dehydrate the air as readily as they will dehydrate the specimen. It is desirable, therefore, either to use fresh absolute alcohol from an unopened bottle or, if one is not using the whole bottle at a time, to keep a layer of some good dehydrating agent at the bottom of the bottle. The best dehydrant for use in absolute alcohol is anhydrous copper sulfate, for this not only absorbs water readily but also indicates, by changing from white to blue, when it is becoming exhausted. Anhydrous calcium sulfate, in the form commercially known as Drierite, is a somewhat better dehydrating agent but cannot be used as an indicator in alcohol.

One may also anticipate that both the hydrocarbons and the essential oils used for clearing will be in a water-saturated condition when purchased. These, therefore, should always be dehydrated as soon as they have been purchased—preferably using Drierite—but, since they have little tendency to absorb moisture from the air, it is not necessary to keep them in bottles containing a dehydrating agent.

CHAPTER 9 | Mounts and Mountants

The final mounting of an object or section for microscopic examination consists in cementing it between a slide and coverslip in a medium or mountant that will preserve it permanently and retain it in a sufficiently transparent condition for study. There are two kinds of mountants: (1) those which are miscible with water and to which objects may be transferred directly and (2) those which are not miscible with water and which require that the object be prepared by dehydration and clearing as described in the last chapter.

Too little attention is paid nowadays to the water-soluble mountants, which are really much more suitable for mounting many objects than the balsam usually employed. Stained objects, of course, cannot be mounted in these aqueous media, but a large number of small invertebrates, particularly arthropods, make better preparations in gum media than they do in balsam.

GUM MEDIA

There are two types of gum media: (1) those of relatively low index of refraction, which do not render the objects placed in them very transparent, and (2) those of a very high index of refraction for use in circumstances where a transparency almost equivalent to a balsam mount is required. The two most useful of the low-refractive-index media are:

Farrants's Medium:

Water	40 ml
Gum acacia	40 g

Glycerin	20 ml
Phenol	0.1 g

The original formula used a saturated solution of arsenous oxide as a preservative, but phenol is just as good for preventing the growth of fungi. This medium is rather difficult to make up, for it is hard to obtain a sample of gum acacia that is not contaminated with pieces of bark, sand, and dirt. A 50 per cent solution of this material, even when diluted with the glycerin, is difficult to filter, and it is better to obtain this medium from a supplier of materials than to make it up oneself. A medium that avoids this difficulty by using synthetic materials is:

Gray and Wess's Medium:

Polyvinyl alcohol	2 g
70% acetone	7 ml
Glycerin	5 ml
Lactic acid	5 ml
Water	10 ml

First make a smooth paste of the dry alcohol with the acetone. Then mix half (5 ml) of the water with the glycerin and lactic acid and stir this into the paste. Add the remaining 5 ml of water drop by drop, stirring constantly. The mixture thus produced is cloudy at first but will become transparent if heated on a water bath for about 10 min.

The advantage of this medium is that it sets fairly rapidly to a tough consistency, so that a slide made with it may be handled within half an hour. A slide made in Farrants's medium may require drying for several days before it is safe to stand it on edge.

The best known of the high-refractive-index watery media is:

Berlese's Medium:

Water	10 ml
Glacial acetic acid	3 ml
Dextrose syrup	5 ml
Gum acacia	8 g
Chloral hydrate	75 g

Mix the water with the acid and the dextrose syrup. Dissolve the gum acacia in this mixture. This will take a week or so, and the material should be stirred at intervals, care being taken not to include too many air bubbles. When solution of the gum acacia is complete, the chloral hydrate is added and stirred to solution.

This mixture suffers from the same disadvantage as Farrants's medium in that it is difficult to obtain clear gum acacia. There is, however, no other medium of such high refractive index that is suitable for mounting small arthropods. The chloral hydrate seems to act as a narcotic, so that a small specimen placed in this medium usually expands into a relaxed condition with all the appendages well displayed for examination. If difficulty is encountered in drying the medium, it is probably due to the very large quantity of chloral hydrate. In those laboratories where the atmosphere is commonly humid, it would be well to reduce the quantity of chloral hydrate to 60 g.

RESINOUS MEDIA

There are two main types of resinous mounting media: (1) those into which the object may be placed directly from alcohol and (2) those for which the object must first be cleared. The former type of medium is commonly referred to as a "neutral" mountant and is used almost exclusively for mounting stained blood films, which are sensitive to the acid that inevitably develops in balsam mounts. The best known of these neutral mountants is a proprietary compound of secret composition known as *euparal*. This may be obtained commercially, but for those who prefer to prepare their own solutions an excellent substitute is:

Mohr and Wehrle's Medium:

Camsal	10 ml
Gum sandarac	40 g
Eucalyptol	20 ml
Dioxane	20 ml
Paraldehyde	10 ml

Camsal is a viscous fluid produced by the mutual solution of equal quantities of camphor and phenyl salicylate (salol). It is well to make up more of this than is required for the preparation of the medium, since it may be used to clear whole objects before mounting in a medium of this type. There is a superstition that a variety of euparal, known as *green euparal*, preserves the color of hematoxylin stains better than the plain material. The formula given above may be turned into an imitation of green euparal by adding as much of a solution of copper oleate in eucalyptol as is needed to produce the required tint.

The most usual resinous mounting medium is Canada balsam. This is the natural exudate of the balsam tree (*Abies balsamea*). Unfortunately, commercial samples are often contaminated with the exudates of other resinous trees, which render the material less suitable for the preparation of microscope mounts. This balsam, like all other balsams, is a solution of a resin in turpentine and contains, in addition, a series of higher-boiling-point hydrocarbons that serve as plasticizers to render it less brittle when it dries. Canada balsam is commercially obtainable in two forms. Natural balsam is the thick sticky material as it comes from the tree and is best used in the preparation of wholemounts. The other form is dried balsam, which has been heated to the extent that the turpentines are driven off, but the natural plasticizers remain. This material, when dissolved in xylene, is used for mounting sections. Dried balsam, unfortunately, is often carelessly prepared by suppliers, who heat it to the point where the natural plasticizers, as well as the turpentines, are driven off. This results in a very brittle compound, and slides made with it are likely to flake off the coverslip a year or two after they have been mounted. Purchased specimens should, therefore, be examined carefully by pressing them with the thumbnail. If, on pressure with the thumbnail, the dried balsam cracks into a powdery material, the sample should be rejected. The desirable consistency is that in which the thumbnail will just manage to mark a piece without any shattering. It is usually worthwhile to prepare one's own dried balsam from the natural balsam by tipping a pound or two into a shallow metal container (a baking dish is excellent) and placing this on a hot plate, taking due precautions against fire. At intervals a small drop should be removed on the end of a glass rod and placed on an ice-cold sheet of metal, where it will harden. The process of evaporation should be stopped when the test specimen shows the desired characteristics. This dried balsam is usually dissolved in the proportion of 60 g of balsam to 40 g of xylene, although some people, including the author, prefer to substitute chloroform for xylene.

A good many attempts have been made to produce a synthetic resin that will have the desirable qualities of balsam without having the disadvantage of becoming yellow or turning acid with age. At the present time none of these is altogether satisfactory. At least to the beginner it is strongly recommended that he confine himself to natural balsam for wholemounts and a solution of dried balsam in a volatile solvent for mounting sections. These compositions are sold under a variety of trade names, and there is little to choose among them. Their only real advantage is that they are water-white. The fact that they are not acid is an advantage with a few stains, mostly blood-staining compositions, but this neutrality is an actual disadvantage in the preservation of many of the compound triple stains. In the experience of the author, many of them do not seem to attach well to the coverslip, and none of them is sufficiently viscous to be used satisfactorily for wholemounts.

CHAPTER 10 | Making Wholemounts

The last few chapters have surveyed a few of the reagents and processes used in the preparation of microscope slides. It is now time to turn in more detail to the preparation of the slides themselves. Those described in this book are prepared in any one of four ways. First there are wholemounts, which, as the name indicates, are mounts either of whole organisms or of parts of organisms that are sufficiently small or transparent to be studied without specialized treatment. Second, described in the next chapter, are smears, in which a thin layer of some fluid intended for study is spread upon a slide and there stained and mounted. Third are squashes, a self-explanatory term. Last are sections, or thin slices cut from an object either too thick or too complex to be studied as a wholemount; the preparation of sections is described in Chapter 12. Wholemounts are the easiest of all slides to prepare and should be the first to be made by the beginner.

TEMPORARY WHOLEMOUNTS

Temporary wholemounts are prepared by every student for the purpose of examining material under a microscope. No object should ever be examined on a slide until the coverslip has been placed over it. The purpose of the coverslip is not only to prevent water from condensing on the lenses but also to provide a flat surface for observation. Temporary wholemounts are usually prepared in water by the simple process of taking a drop of the fluid containing the material to be examined, such as living paramecium, and lowering the coverslip on the surface. This is perfectly satisfactory, providing it is not necessary to observe the objects for long periods. After a time, however, the water evaporates from the edges of the coverslip, which crushes the material. This may be avoided readily by slowing the evaporation with petroleum jelly, a little ring or square of which is built up in the center of the slide before the drop is placed in position. The coverslip is pressed down very gently until it is seen to be sticking to the petroleum jelly.

There are many tricks that may be used to examine animals that move too rapidly for study. One of the simplest of these is to place a piece of lens paper on the slide and then to place a drop of the culture under examination on this. The addition of a coverslip causes the fibers of the lens paper to make, as it were, a series of little compartments in which the animals become trapped. Another useful device is to mix thoroughly a culture of the animal in question with an equal volume of 0.5 per cent agar. This thickens the liquid sufficiently to slow down most forms. For the same purpose, 0.1 per cent carboxymethyl cellulose may be used.

The term wholemount, however, usually means a permanent preparation made in a medium that will both preserve the object and hold the coverslip in place. These mounting media are either water-miscible, in which case most objects can be placed in them directly, or resinous, requiring extensive preparation of the object. Each type will be described in its turn.

MOUNTING IN GUM MEDIA

The simplest preparations are those made in water-miscible mountants, which are of far wider utility than is usually realized due to a complete mental block on the part of most microscopists when faced with any mounting medium that is not a solution of a resin in a hydrocarbon. As a matter of fact, most simple objects, such as the scales of fish and animal hairs, may be mounted more readily in aqueous than in resinous media. This process of mounting is so simple that it is regarded with distrust by those who have come to believe that only through complexity can good results be produced. With aqueous media the object to be mounted is merely placed in a drop of mountant on the slide, and a coverslip is pressed on top. This process, moreover, is not confined to relatively hard objects but may be applied to many protozoa and other small invertebrates. Small invertebrates do not always make satisfactory permanent mounts by this method, for they ultimately reach a refractive index identical with that of the mountants and thus vanish. A temporary mount of a paramecium in one of these media, however, will show the internal structure better than will the average stained mount, while it will also give a clearer indication of what the living object looked like. The most common objects to be mounted by this method are small arthropods; a description of the preparation of a mite in one of these media is given in Part Three.

Finishing Slides in Gum Media. Slides may be left exactly as they are prepared, but this will give a rather clumsy appearance, since some of the mountant will exude from under the coverslip. This exudate may be removed by washing with warm water, but it will be some time after this before the gum at the edges of the coverslip is dried. Moreover, no mounting medium containing glycerin can fail to absorb moisture from the air on humid days and to lose it on dry days, so that it is usually better to finish the slide by applying a ring of varnish around the outside.

Preparation of Thick Objects. Large objects do not usually make good preparations in gum media because they take a very long time to become transparent. It is better, therefore, to mount them in resinous media in the manner about to be described; but, before doing so, it is necessary to explain some of the methods whereby a coverslip may be held in place over a relatively large form. It is obvious that unless support is provided for the coverslip, it is bound to tip to one side or the other and thus make the mount relatively useless.

A good method of supporting a coverslip over a large object is to cut sheets of celluloid of various thicknesses into little squares spaced equally around the edge before the coverslip is applied. These squares may be made of any desired thickness, to accommodate objects of varying size. This method, however, is not satisfactory for objects much more than 0.5 mm thick because the wide rim left at the edge of the cover causes the mountant to dry out very rapidly.

An alternative is to provide a little box, known as a *cell*, and to cement this to the slide before mounting the object in it. Cells are available in a great variety of materials, and their use is described in some detail in Gray's "Microtomist's Formulary and Guide." For elementary purposes, the best cells that can be used are undoubtedly stamped from sheet tin or pewter and may be obtained in thicknesses varying from ½ to about 1½ mm. They should be cemented to the slide with some cement that is not soluble in balsam. The best general-purpose adhesive is gold size; reference should be made to the source cited above for details of this material and the method of its application. Glass cells are available from biological supply houses. Although they look very attractive, there is no real advantage in their use since the object is usually studied by transmitted light.

MOUNTING IN RESINOUS MEDIA

Resinous media are used for wholemounts not only because they permit the mounting of stained objects but more particularly because they make the specimen more transparent. This transparency comes from the increase in the index of refraction of the specimen when it is completely impregnated with the resin. These resins, however, are not miscible with water, so that one is forced first to remove the water (dehydration) and then to replace the dehydrant with some material (clearing agent) with which the resin itself is miscible. Before these operations are conducted, the specimen must be killed and hardened (fixed), and it is customary to stain the specimen in order to bring out those internal structures which would become invisible, were they not colored, through the increase in transparency. All the following operations, therefore, must be executed, and each will be discussed in turn:

1. Narcotizing and fixing—to preserve the shape
2. Staining—to bring out structure
3. Dehydrating—to remove water
4. Clearing—to permit impregnation with resin
5. Mounting in resin—to preserve in transparent form

Narcotizing and Fixing Specimens. Hard objects, such as small arthropods and hairs, may be dehydrated and mounted directly in resinous media, but they are better prepared in gum media. Most objects that are mounted in resinous media are too soft to withstand the process of dehydration and clearing without special treatment. Although hardening and fixing agents were once considered separate entities, they are now usually combined into a solution known as a *fixative*. Some of the more useful of these solutions are given in Chapter 6. Few small animals, when plunged into a fixative, will retain their shape, so that it is necessary first to narcotize them in some solution that will render them incapable of muscular contraction.

Narcotization should always proceed slowly; that is, a quantity of narcotic should be added at the beginning and increased later, adding the fixative only after cessation of movement. This is easy to judge in the case of motile forms, which may be presumed to be narcotized shortly after they have fallen to the bottom of the container, but in the case of sessile forms it is necessary to use a fine probe, preferably a hair, to determine the end point of narcotization. Some recommended narcotic mixtures are given at the end of Chapter 6. The following suggestions for various types of invertebrates should be used as a basis for further experiment.

Noncontractile Protozoa.   These do not require narcotization and may be fixed directly in a chrome-acetic mixture.

Individual Contractile Protozoa.   These are very difficult to handle. Individual rhizopods, such as ameba, are best fixed to a coverslip in the following manner. Take a clean coverslip and smear on it a very slight quantity of fresh egg white. Place each individual protozoan in the center of the coverslip and allow it to expand. While this is going on, fit a flask or kettle with a cork through which is inserted a glass tube. The outer end of the tube should be drawn to a fairly fine point. Boil the water in the flask to produce a jet of steam. As soon as the animal is satisfactorily expanded, pick up the coverslip very gently and pass the underside momentarily through the jet of steam. This instantly hardens the protozoan in position and at the same time cements it to the coverslip through the coagulation of the egg white. Then transfer the coverslip to any standard fixative solution for a few minutes before washing and storing in alcohol.

Coelenterata.   Hydroids are usually narcotized with menthol, although the author prefers his own mixture for the purpose. They are best fixed in a hot mercuric-acetic mixture.

Platyhelminthes.   Some of the smaller fresh-water Turbellaria (e.g., Vortex, Microstomum) may be narcotized satisfactorily by adding small quantities of 2 per cent chloral hydrate to the water in which they are swimming. Another good technique is to isolate the forms in a watch glass of water and place the watch glass under a bell jar together with a small beaker of ether. The ether vapor dissolves in the water and narcotizes these forms excellently. A detailed account of the method of handling the liver fluke is given in Example 4 and may be employed satisfactorily for other parasitic flatworms.

Annelida.   Small marine free-living Polychaeta make excellent wholemounts and do not usually require narcotization before killing. They should, however, be stranded on a slide, and a small quantity of the fixative should be dropped on them so that they die in the flat condition that makes subsequent mounting possible. Much more realistic mounts are obtained by this means than if the animals are laboriously straightened before fixing because they usually contract into the sinuous wave that they show when swimming. There seems to be no certain method of fixing the Nereids with their jaws protruding.

Fresh-water Oligochaetes. These are best narcotized with chloroform, either by adding small quantities of a saturated solution of chloroform in water or by placing them in a small quantity of water under a bell jar in which an atmosphere of chloroform vapor is maintained. Leeches are rather difficult to handle, and the author has had greatest success by placing them in a large quantity of water to which is added, from time to time, small quantities of a solution of magnesium sulfate. As soon as the leeches have fallen to the bottom, much larger quantities of magnesium sulfate can be added, which will leave the leeches, in a short time, in a perfectly relaxed but not expanded condition. Then they should be flattened between two slides and fixed in Zenker's fluid. After the specimens have been fixed sufficiently long to hold their shape when the glass plates are removed, they are transferred for a couple of days to fresh fixative and then washed in running water overnight.

Bryozoa. Marine Bryozoa may be narcotized without the least difficulty by sprinkling menthol on the surface of the water containing them. Subsequent fixation is best in some chrome-acetic mixture. It is usually recommended that fresh-water Bryozoa be narcotized in a cocaine solution, but the author has found menthol just as good and very much easier to use. Fresh-water Bryozoa should be fixed directly in 4 per cent formaldehyde since they shrink badly in any other fixative.

Arthropods. Wholemounts of most small arthropods are better made in gum media.

Choice of a Stain. Whatever method of narcotization and fixation has been employed, the specimens that are to be mounted are washed free of fixative and accumulated in either water or 70 per cent alcohol.

Small Invertebrates and Invertebrate Larvae. These are best stained in carmine by the indirect process, that is, by overstaining and subsequent differentiation in acid alcohol. For most specimens the author prefers Grenacher's alcoholic borax carmine.

Larger Invertebrate Specimens. Larger specimens are better stained by the direct process, that is, exposed for a considerable length of time to a very weak solution of stain and not differentiated.

Vertebrate Embryos. These seem to stain more satisfactorily in hematoxylin than in carmine solutions. The author's preference is the formula of Delafield. Detailed instructions for the use of this stain on a chicken embryo are given in Example 3.

Plant Materials. Plant specimens that are to be prepared as wholemounts often consist of only one or two layers of cells and are therefore easier to stain than zoological specimens. The nuclei may be stained either with safranin or with an iron hematoxylin technique which in zoological procedures is rigorously confined to sections. A contrasting plasma stain may be used after the nuclei have been well differentiated.

Dehydration. The specimens, plant or animal, stained or unstained, are accumulated either in distilled water or in 70 per cent alcohol according to the treatment they have had. It is necessary to remove the water from them before they can be mounted in any resinous medium. Ethyl alcohol is widely used as a dehydrant, and, at least in the preparation of wholemounts, only its unavailability should make any substitute necessary. Where substitution is necessary, acetone or methyl alcohol, in that order of preference, may be used. Both have the disadvantage of being more volatile than ethyl alcohol and, therefore, requiring more care in handling.

Dehydration of animal material is carried out by soaking the specimen in gradually increasing strengths of alcohol; it is conventional to employ 30 per cent, 50 per cent, 70 per cent, 90 per cent, 95 per cent, and absolute alcohol. The author prefers to omit from this series, unless the object is very delicate, both the 30 and the 50 per cent alcohol, thus starting with direct transfer from water to 70 per cent alcohol. The only difficulty likely to be met in dehydration is in the handling of small specimens because, if they are in specimen tubes, it is almost impossible to transfer them from one to the other without carrying over too much

Fig. 10-1 | Transferring objects between reagents with cloth-bottomed tubes.

weak alcohol. The author has long since abandoned tubes in favor of the device shown in Fig. 10-1. This is a short length of glass tube, open at both ends, which has a small piece of bolting silk or other fine cloth tied across the lower end. The specimens are placed in the tubes which, as may be seen in the illustration, are transferred from one stender dish to another with a minimum chance of contamination. These tubes are commercially available in England, but in America they must be imported or homemade.

There is no means of judging when dehydration is complete save to attempt to clear the object. It is unwise to believe the label on an open bottle or jar if it says "absolute alcohol" because this reagent is hygroscopic and rapidly absorbs water in the air. A quantity of anhydrous copper sulfate should therefore be kept at the bottom of the absolute alcohol bottle and the alcohol no longer regarded as absolute when the salt starts turning from white to blue. More wholemounts are ruined by being imperfectly dehydrated than by any other method. Even the smallest specimen should have at least 24 hr in absolute alcohol before any attempt is made to clear it.

Choice of a Clearing Agent. A clearing agent must be miscible both with absolute alcohol and with the resinous medium that has been selected for mounting. The ideal substances for this purpose are essential oils. They impart just as much transparency to the specimen as the resin used for mounting, so that they provide, as it were, a preview of the finished specimen. The use of benzene, which is recommended for preparation of paraffin sections, has started to spread into the preparations of wholemounts. In the author's opinion, it is utterly worthless for this purpose. It has a relatively low index of refraction, so that it is impossible to determine whether or not the slight cloudiness of the specimen is the result of imperfect dehydration until after the specimen has been mounted in balsam.

The first choice is terpineol (synthetic oil of lilac), which has advantages possessed by no other oil. It is readily miscible with 90 per cent alcohol, so that it will remove from the specimen any traces of water that may remain in it through faulty dehydration, and it does not make specimens brittle. It also has a very slight and rather pleasant odor. Clove oil is the most widely recommended essential oil for the preparation of wholemounts. It has only two disadvantages: its violent odor and the fact that objects placed in it are rendered brittle. If a small arthropod is cleared in clove oil, it is almost impossible to get the animal into a wholemount without breaking off some appendages. Clove oil, however, is miscible with 90 per cent alcohol.

Mounting in Balsam. Nothing is easier than to mount a specimen in balsam, provided that it has been perfectly dehydrated and cleared. A properly made wholemount should be glass-clear, but it will not be clear in balsam unless it is clear in terpineol or clove oil. Not more than one in a thousand wholemounts has this vitreous appearance. The worker who is accustomed to looking at rather cloudy wholemounts should take the trouble to dehydrate a specimen thoroughly, to remove the whole of the dehydrating agent with a clearing agent, and then to mount the specimen properly in balsam.

The first step in making a mount in Canada balsam is therefore to make quite certain that the specimen in its essential oil is glass-clear. The second step is to use natural Canada balsam and not dried balsam dissolved in xylene. Solutions of dried balsam in hydrocarbons are meant for mounting sections and, for this purpose, are superior to the natural balsam.

Fig. 10-2 | **Applying coverslip to balsam wholemount.** Notice that the coverslip is held horizontally centered over the drop of balsam.

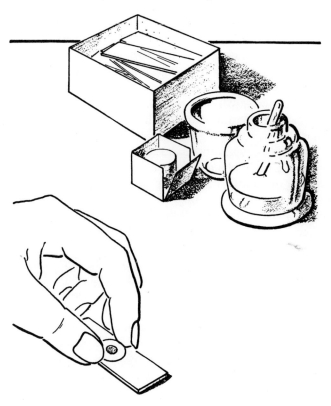

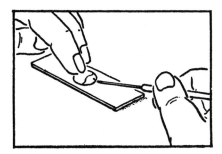

Fig. 10-3 | **Wrong way to apply coverslip to balsam wholemount.** This method draws the object to one side.

Natural balsam is, however, just as suitable for wholemounts and is just as easy to obtain. If it is found to be too thick for ready use, it may be warmed gently to the desired consistency. A single small specimen is mounted by placing it in a drop of balsam on a slide and then lowering a coverslip horizontally (Fig. 10-2) until the central portion touches the drop. The coverslip is released and pressed very gently until it just touches the top of the object. By this means it is possible to retain the object in the center of the coverslip and also, when using natural balsam, which does not shrink in drying, to avoid cells for any but the largest objects. Unfortunately most people are accustomed to mounting sections in thin balsam by the technique shown in Fig. 10-3, that is, by touching one edge of the coverslip to the drop and then lowering it from one side. The objection to this is that the balsam, as is seen in Fig. 10-3, immediately runs into the angle of the coverslip, taking the object with it, and it is difficult to lower the coverslip in such a way that the object is left in the center. When mounting specimens or deep objects in a cell in which a cavity has been ground, it is desirable to hold the coverslip in place with a clip while the balsam is hardening. This process is seen in Fig. 10-4; the type of clip there shown is made of phosphor-bronze wire and is far superior to any other type.

The description above presumes the use of natural Canada balsam, unquestionably the best resinous medium in which to prepare wholemounts. If a solution of dried balsam in xylene is used, a very different technique will have to be adopted. In the first place, most of these solutions are so thin that it is almost impossible to apply the coverslip as shown in Fig. 10-2, and the technique shown in Fig. 10-3 must be adopted. This difficulty may be avoided by placing the object on the slide, putting a drop of the medium over the object, and then setting the slide in a desiccator until most of the solvent has evaporated. A second layer is placed on top to build up a large drop or, rather, a thick coat of varnish over the specimen. A coverslip is applied, and the slide is warmed until the resin becomes fluid.

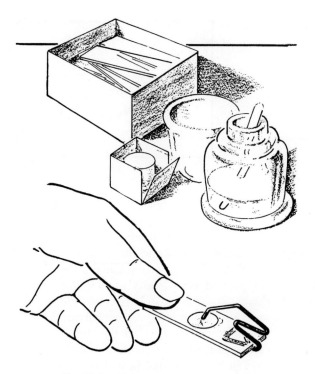

Fig. 10-4 | **Balsam wholemount ready for drying.** The phosphor-bronze clip prevents the coverslip from moving.

The best use for solutions of balsam in making wholemounts is in dealing with a very large number of objects. The method in this case is to transfer the objects from their clearing medium to a tube or dish of the solution of balsam in whatever hydrocarbon has been selected and then permit the hydrocarbon to evaporate. When the balsam that remains has reached a good consistency for mounting, take each specimen, together with a drop of balsam, place it on a slide, and add a coverslip. By this method large numbers of slides can be made in a short time. It is not necessary to use solutions of dried balsam, and the author prefers, for this purpose, to dilute natural balsam with benzene. Mounting large objects in a deep cell in Canada balsam is not to be recommended because the balsam becomes yellow with age and, when in thick layers, tends to obscure the specimen. A wholemount of a 96-hr chicken embryo, for example, is of extremely doubtful value, but if it has to be made it is best first to impregnate it thoroughly with a fairly thin dilution of natural balsam. It is then placed in the cell, piling the solution up on top, and left in a desiccator. The cell is refilled as the evaporation diminishes the contents; when completely filled with solvent-free balsam, it is warmed on a hot table. The coverslip is applied directly.

Finishing Balsam Mounts. If a mount has been made correctly with natural balsam, and if the size of the drop has been estimated correctly, no finishing is required since no balsam will overflow the edges of the coverslip. Natural balsam is very thick when cold, so that the coverslip will not become displaced if the mount is handled before the balsam is fully hard. The hardening may be aided by heat, but care must be exercised in heating thick mounts, particularly where the coverslip is not supported by a cell, that the liquefaction of the balsam does not cause the coverslip to tip sideways. Despite the fact that drying time is sometimes prolonged, natural balsam should always be used for thick mounts because, if a solution of dried balsam and xylene is employed, the evaporation of the solution will cause huge air bubbles to be drawn under the coverslip. When it is sufficiently hard, the slide should be cleaned, first, by chipping off any excess balsam with a knife and, secondly, by wiping away any chips with a rag moistened in 90 per cent alcohol. This will leave a whitish film over the surface of the slide, which may then be removed with a warm soap solution. The slide should be polished before being labeled.

With regard to labeling, it may be pointed out that no power on earth will persuade gum arabic, customarily used for attaching labels, to adhere to a greasy or oily slide. The portion of the slide to which the label is to be attached should therefore be cleaned more carefully than any other. The author prefers to moisten both sides of the label, press it firmly to the glass, and write on it only after it is dry. The so-called "self-adhesive" labels should be avoided for other than temporary purposes. The chlorinated rubber which is the basis of most of these adhesives oxidizes over a period of time.

SUGGESTED ADDITIONAL READING

Gray, P.: "The Microtomist's Formulary and Guide," New York, McGraw-Hill Book Company, 1954.

CHAPTER 11 | Making Smears and Squashes

The last chapter was concerned with the preparation of microscope slides from whole objects preserved as nearly as possible in their natural shape; the next chapter will be concerned with the preparation of thin slices of objects, or sections. Between the extremes of a whole object and a thin slice, there are the types of preparation that are discussed in this chapter. These are the smear and the squash, which are exactly what their names indicate; that is, they are prepared by smearing or squashing some substance on a clean glass slide where it may be fixed, stained, and mounted. Three operations are necessary in the preparation of smears of fluids: (1) the smearing of the material itself into a layer of the required thickness; (2) the fixing of this layer both to ensure its adherence to the slide and to make sure that the contained cells remain in their normal shape; and (3) the staining and mounting of the fixed smear. Each of these operations will be discussed successively.

SMEARS

Preparation of the Smear. The first thing to do in the preparation of a smear is to make sure that there are some chemically clean slides available. Only materials containing large quantities of protein, such as blood, will adhere to slides that are not perfectly clean. Any method may be used for cleaning slides. For this particular purpose, however, the author prefers to use any household scouring powder, which consists of a soft abrasive together with some detergent agent. The powder is made into a thin cream with water. Each slide is then dipped into the cream and put in a rack to dry. As soon as it has dried, the slide may be returned to a box, preferably separated from the next slide with a thin paper insert. Since slides are commonly sold with paper separators, they may be stored with the separators in the original box.

Two or three hundred slides can be prepared easily and quickly in this manner and stored for future use. When a slide is needed, the white powder is polished from the surface with a clean linen or silk cloth. Smears often have to be made at unexpected moments, so that it is a great convenience to have slides at hand that may be rendered fit for use in a few moments.

The actual method of smearing varies greatly according to the material being used. Probably more smears are made of blood than of any other fluid, and the technique for the preparation of these is so well established that it will be described as a type. The material itself may be either taken from the puncture wound directly onto the slide or, as shown in Fig. 11-1, removed from the puncture wound with a pipette and transferred to the slide. The drop is placed about 1 in. from one end of the slide. A second slide, as shown in Fig. 11-2, is placed in front of the drop. Capillary attraction will distribute the fluid along the edge of the second slide, which is then (Fig. 11-3) pushed sharply forward until it reaches the end of the bottom slide. The material of which the smear is made is thus dragged out behind the first slide and distributed more or less uniformly on the under slide. A few people still try to conduct the operation in the reverse manner by placing the second slide on top of the first, sloping it at a reverse angle to that shown, and then

Fig. 11-1 | Making a smear preparation. Place the drop about 1 in. from the end of the slide.

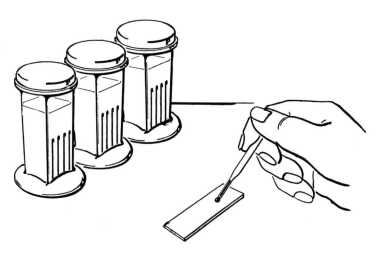

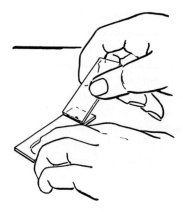

Fig. 11-2 | Making a smear preparation.
Apply a second slide just in front of the drop.

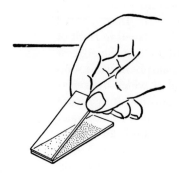

Fig. 11-3 | Making a smear preparation.
Push the slide smoothly forward to spread the smear.

endeavoring to push rather than drag the material across the lower slide. The objection to this is that it results in crushing cells, although it must be admitted that it frequently results in a more uniform distribution of the material.

The method just described is the standard procedure for producing thin smears. These are necessary for those fluids, such as vertebrate blood or mammalian seminal fluid, which contain large numbers of objects requiring wide separation for satisfactory study.

There are a number of fluids, however, from which thick smears must be made, either because they contain relatively few cells, as in the case of invertebrate blood, or because they contain parasites that are distributed relatively sparsely through the material, as in the case of malarial diagnostic smears. These thick smears are commonly made with the aid of a loop of wire held in a needle holder of the type found in bacteriological laboratories. This loop is dipped into the fluid to be examined.

The material is spread with a rotary motion in the center of the slide. This is very similar to the preparation of smears of bacterial material, which is described in some detail in Example 7.

Fixing Smears. Smears may be fixed by drying, by alcohol, or in one of the conventional fixatives. When a smear is to be fixed only by drying, it is dried by waving it in the air, as soon as it has been made, and then setting it aside for later treatment. This procedure is excellent in the case of objects such as bacteria or erythrocytes, which do not change their shape after drying, or for materials such as white blood corpuscles, which it is not desired to preserve in their normal shape. No other object, however, can be considered satisfactory unless it has been fixed. The simplest method of doing this is to pass the smear, just as it is drying, through a jet of steam. This technique has already been described for mounting amebas and need not be repeated here.

All other smears should be fixed before they are dried, and it is something of a problem to fix them without removing the material from the slide. It is obvious that if the material is freshly smeared onto a glass slide and then dropped into a fixative of some kind or another, it will be likely to be washed off. The logical solution to the problem is to use a fixative in a vapor phase, and nothing is better for this purpose than osmic acid. To use this material, take a petri dish and place in it a couple of thin glass rods sufficiently far apart to permit the slide to rest on them without the smear touching them. Then place a drop or two of a solution of osmic acid, usually of 2 per cent strength, in the bottom of the petri dish and replace the cover. It must be emphasized that osmic acid fixes the mucous membrane of the nose and throat just as readily as it does a smear, and every caution must be taken to avoid inhaling the vapors of this material. As soon as the smear is made *and before it has time to dry*, it is placed face down across the two glass rods, so that it is exposed to the vapor but not to the liquid. The cover is then replaced on the petri dish and the slide left in place for 3 to 4 min, in the case of a thin smear, or 5 to 10 min in the case of a thick one. Then it is transferred to distilled water to await staining.

It occasionally happens that one must fix a slide in one of the conventional fluid fixatives. This is done with the same petri dish and glass rod setup as is used for vapor fixation, but in this instance the fixative is carefully poured into the petri dish, which must be level, until it has reached a depth such that when the slide is laid across the glass rods, the underside of the slide with the smear on it is in contact with the fluid, while the upper part is free from fluid. If the smear is reasonably thin and is laid carefully in place, it usually will not become detached.

Staining Smears. Blood smears are stained so universally with one or another of the methylene blue–eosinate mixtures that it comes as something of a surprise to most people to learn that any stain that is suitable for sections may also be employed for smears. The advantage of these mixtures for blood films is that the solvent methyl alcohol acts as a fixative, so that the films, in effect, are stained and fixed in the same operation. Where a blood smear is to be used for diagnostic purposes, these techniques are excellent, since the appearance of the various types of white corpuscles under this treatment is known to every technician. For materials other than blood, there is no limit to the type of staining that may be employed, although it must be remembered that these very thin films require a stain of considerable intensity if the finer structures are to be seen.

SQUASHES

Squashes are exactly what the name indicates and no theoretical discussion of this process is necessary. The object is placed on a slide, crushed under a coverslip, and examined.

The choice of the fluid in which it is crushed depends on the nature of the object and the purpose of the preparation. Hydra, for example, may be crushed in water—although cold-blooded Ringer's solution (page 230) is better—for the purpose of displaying the nematocysts. The anthers of an embryo flower or the salivary glands of a Drosophila larva may be squashed in a drop of stain to facilitate the study of the chromosomes.

Squashes are usually regarded as temporary preparations and are undoubtedly best used for this purpose. The difficulty in transforming them into permanent mounts is that the removal of the coverslip, to permit dehydration and clearing, often disturbs or even removes the squash itself. This can often be prevented by freezing. In this process the slide is laid on a block of dry ice—the stage of a freezing microtome is less effective—and left for about 5 min, or until the coverslip is thoroughly coated with frost. A safety-razor blade can then be gently slid along the slide and the coverslip levered off the preparation. The adhesion of the latter to the slide through the subsequent processes of dehydration and clearing is, of course, dependent on the nature of the preparation.

CHAPTER 12 | Making Sections

NATURE OF THE PROCESS

A section is a thin slice cut from biological materials with a view to studying either the cells themselves or their arrangement, neither of which can be made out from a wholemount.

Though sections may be cut at any angle, they are usually taken through any one of three planes (see Fig. 12-1), which are known as *transverse, sagittal,* and *frontal* planes. The purpose of this orienta-

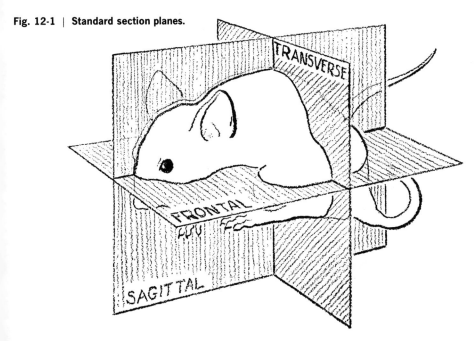

Fig. 12-1 | Standard section planes.

tion of the material is a better realization of the structure of the whole from an examination of the section. Theoretically, to produce a section, it is necessary only to take a sharp knife and cut a thin slice from the object under examination. Very few materials, however, are suitable for this, and this procedure does not produce sections of the same thickness. It is therefore customary to use an instrument known as a *microtome*, which is a device for advancing a block of tissue a given amount, cutting a slice from it, readvancing it the same amount, and repeating the process.

Another objection to the mere cutting of slices from an object is the nature of biological specimens themselves. Very few of these are stiff enough to withstand the action of the knife without bending, and many contain cavities that would be crushed out of recognition as the section was taken. It is customary in most biological work, therefore, to surround and support the object to be cut with some material that will impregnate its whole substance. The medium most commonly used to support structures is wax. The technique for cutting wax sections is described later. There are, however, a number of materials that may be cut without either complicated microtomes or the support of impregnating substances. Sections which are so cut are known as *free*, or *freehand*, sections.

FREE SECTIONS

Microtome for Free Sections. Even if the material itself is of the correct consistency to withstand the action of the knife, it is still necessary to have some mechanism that will allow the production of sections of known thickness. The type of microtome usually employed in hand

Fig. 12-2 | **Hand microtome.**

sectioning is shown in Fig. 12-2 and consists essentially of a disk, usually of highly polished plate glass, supported on a cylinder, which is gripped in the hand. For holding specimens within this cylinder there is a mechanism that terminates at its lower end on a micrometer screw. When this screw is turned, therefore, the object in the holder is pushed above the surface of the glass plate. The collar of the micrometer screw is graduated, sometimes in thousandths of an inch, but more usually in hundredths of a millimeter. The unit commonly employed to describe the thickness of a section is a micron (μ), which is one-thousandth of a millimeter, but hand sections are rarely cut less than 10 μ thick and are usually better at two or three times this.

Methods of Holding Material. The material, although it may be suitable for cutting, is rarely of a size and shape that may be gripped in the holder of the hand microtome without additional support. Therefore, it must be held in some substance that itself will cut readily and that may be shaped easily to support the material to be cut. It is perfectly possible to embed the material in wax before cutting a hand section, but if one is to go to this amount of trouble it is usually better to employ a complex microtome of the type described later. Vegetable tissues are generally used to support objects for hand sectioning. The two best known are elder pith and carrots. Elder pith has the advantage that it may be stored indefinitely and cuts with a clean crisp action. Unfortunately, the pith of the American elder (*Sambucus canadensis*) does not appear to be so suitable for the purpose as the pith of the European elder (*S. nigra*). This difference between the two species may account for the disfavor in which elder pith is held in the United States, for in the author's experience it is far more convenient than the carrot. The disadvantage of the carrot is that it must be absolutely fresh, and, even if it is kept in water overnight, it loses much of that crispness necessary for the production of a good section.

Almost all hand sections are cut from botanical material, the majority of them from leaves or stems. To support a leaf, merely cut a cylinder of the right diameter to fit in the microtome from either elder pith or carrot, split it down the middle, insert the leaf (Fig. 12-3), and then tighten the holder. Stems, however, cannot be held by this means, so that one must obtain a hollow cylinder, having an outer diameter convenient to the microtome being employed and an inner diameter slightly less than that of the stem to be gripped. This hollow cylinder is split, the stem inserted, and the section cut (Fig. 12-4). Of course, a few substances, such as cork or stiff plant stems, may be cut without any other support; these are, however, in the minority.

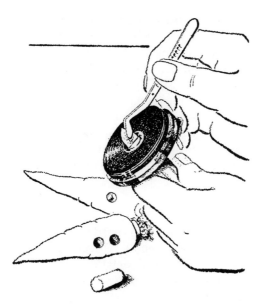

Fig. 12-3 | Inserting a leaf into a split cylinder of carrot.

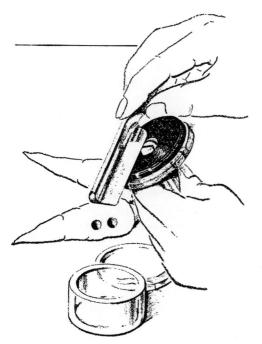

Fig. 12-4 | Cutting a hand section. The razor is drawn across the plate with gentle pressure, and the section is then washed into a stender dish.

Hardening and Fixing Materials for Cutting. Many objects that are in themselves unsuitable for sectioning by hand can be rendered more suitable if they are fixed and hardened in some chemical reagent. If, however, one is to go to the trouble of hardening and fixing material in a formula designed to preserve the structure of the cells, it is usually worthwhile going to the additional trouble of embedding the material and cutting paraffin sections. It is generally sufficient for material that is to be hand sectioned to be preserved in 90 per cent alcohol, a process that is equally applicable to the stems and leaves of botanical specimens and to the very few animal materials, such as cartilage, which are suitable for the production of hand sections.

Staining and Mounting Hand Sections. Sections are taken from the knife as individual objects and are accumulated in a dish of some preservative, usually 70 per cent alcohol. They should be treated as wholemounts rather than as sections. That is, either they may be mounted directly in gum media or they may be stained and mounted in resinous media in the manner described in Chapter 6.

PARAFFIN SECTIONS

Preparation of paraffin sections is quite a complex operation and involves the following stages:

1. Fixation of the material
2. Dehydration in order that the material may be impregnated with a fluid capable of dissolving wax
3. Removal of the dehydrating agent with a material solvent of, or miscible with, molten wax
4. Soaking the cleared specimen in molten wax long enough to ensure that it will become completely impregnated
5. Casting the now impregnated specimen into a rectangular block of wax
6. Attaching this block of wax to some holder, which itself may be inserted into a suitable microtome
7. Actual cutting of the block into ribbons of sections
8. Placing these ribbons on a glass slide in such a manner that they will lie flat and that the contained section will be adherent after the wax has been dissolved
9. Removal of the wax solvent
10. Staining and mounting

Each of these operations will be dealt with in due order. Part Three consists of a series of examples which describe in detail the application of these principles to actual preparations.

Choice of Fixative. The methods described in Chapter 10 for the fixation of objects for wholemounting can be used equally well if these objects are to be sectioned. The selection of the fixative for blocks of tissue, however, is based more on the nature of the detail that is to be preserved. In general, it may be said that strongly acid fixatives are best where nuclear detail is required. The selection of fixatives and several hundred formulas for the solutions involved are given in Gray's "Microtomist's Formulary and Guide," to which reference should be made by the student seeking special information. For elementary histological preparations, the fluids of Zenker, Gilson, and Petrunkewitsch, the formulas for which are given in Chapter 6, are all excellent.

Dehydrating. The classic method of dehydration is to soak the object in a graded series of alcohols, usually 10 or 15 per cent apart. Dehydration through gradually increasing strengths of alcohol may be vital when one is dealing with delicate objects containing easily collapsible cavities, such as chick and pig embryos, but a block of tissue may be taken from water to 96 per cent alcohol without any appreciable damage. Even if one uses increasing strengths of alcohol, the series normally in employment at the present time is by no means satisfactory. It is customary, for example, to pass the object from water to 30 per cent alcohol at one end of the series and to pass it from 85 per cent to 96 per cent at the other end of the series. An intelligently graded series for delicate objects should run from water to 15 per cent alcohol to 40 per cent alcohol to 75 per cent alcohol to 96 per cent alcohol rather than through the conventionally spaced gradations. This is not at all in accordance with the recommendations in most textbooks but is based on the author's experience over a long period of time. In using this classic method of dehydration, it is not necessary to confine the technique to ethyl alcohol. Methyl alcohol or acetone will dehydrate just as effectively, although they are more volatile.

The substitution of a solvent that is miscible both with water and molten wax for a straight dehydrating agent is in vogue today. The best known of these solvents is dioxane, though n-butyl alcohol has also been recommended. The author is not completely satisfied with these methods because, although the solvents involved are excellent dehydrating agents, they are relatively poor solvents of paraffin and frequently occasion great shrinkage of delicate objects in the final transition between the solvent and the wax. For such purposes as the routine examination of the tissue blocks in a pathological laboratory or the sectioning of relatively sturdy plant materials, they may justifiably be employed. However, for sections in which structures are to be retained intact for subsequent research, it is to be recommended most strongly that the standard routine of passing from a dehydrating to a clearing agent be followed.

Clearing. The choice of a clearing agent in section cutting is of far more importance than the choice of a dehydrant, since there is not the slightest doubt that prolonged immersion in xylene leads to a hardening of the tissue with subsequent difficulty in sectioning. Benzene is much to be preferred for most objects.

It is still recommended occasionally that essential oils, such as cedar oil, be used for clearing objects for embedding. There is no justification for this unless it is vital that the object be rendered transparent rather than alcohol-free, in order that some feature of its internal anatomy may be oriented in relation to the knife. Essential oils are excellent for wholemounts, but they are not readily removed from the specimen by molten wax, so that if they must be used they should always be washed out with a hydrocarbon before the wax bath. Relatively small traces of any essential oil will destroy the excellent cutting properties of any wax mixture, and, as the oils are nonvolatile, there is no chance of getting rid of them in the embedding oven.

Dehydrating and Clearing Plant Tissues. As was pointed out in Chapter 8, the methods of clearing and dehydrating animal tissues cannot satisfactorily be applied to plant tissues because they cause the cell contents to pull away from the cell walls. Many methods have been suggested, but the majority of botanists today prefer the Zirkle technique. A schedule of this type, which has been found satisfactory but which workers may wish to modify from their own experience, involves passing the tissues through the following mixtures:

1. Water 95
   Alcohol 5
2. Water 90
   Alcohol 10
3. Water 82
   Alcohol 18
4. Water 70
   Alcohol 30
5. Water 50
   Alcohol 40
   Tertiary butyl alcohol 10
6. Water 30
   Alcohol 50
   Tertiary butyl alcohol 20
7. Water 15
   Alcohol 50
   Tertiary butyl alcohol 35
8. Water 5
   Alcohol 40
   Tertiary butyl alcohol 55
9. Alcohol 25
   Tertiary butyl alcohol 75
10. Tertiary butyl alcohol

The length of time that the tissues remain in each stage depends on their size but should be not less than 1 hr nor more than 6 hr in the first five fluids. The sixth mixture is the critical point, and most people leave even small tissues in it overnight. One hour in each of the remaining fluids is usually enough for medium-sized pieces.

Choice of an Embedding Medium. It is to be presumed at the present time that a mixture will be used instead of a plain paraffin. If a plain paraffin is preferred, then it is necessary to buy (in the United States by importation) a carefully fractionated and very expensive wax. Ordinary cheap paraffin is a mixture of a great variety of compounds of slightly different melting points; it is essential in the use of pure wax that a wax of a very sharp melting point be obtained.

The choice of an embedding medium should be dictated less by the nature of the specimen than by the conditions under which it should be cut. If pure paraffin is to be employed, it should be of such a melting point as will give the hardened wax a crisp section at the required room temperature. Since the introduction of any foreign substance automatically lowers the melting point of the wax, it is obviously desirable to use mixtures rather than the pure material. For ordinary routine preparations, the author's preference is for any of the paraffin–rubber–bayberry wax mixtures. The introduction of rubber undoubtedly increases the stickiness of the wax and makes it easier to obtain continuous ribbons. Bayberry wax not only prevents the crystallization of the paraffin but also lowers its melting point.

There are many proprietary mixtures available on the market. For those who prefer to prepare their own, the two following compositions are excellent:

Hance's Rubber Paraffin:

Stock rubber solution
    Cut 20 g of crude rubber into small pieces and dissolve, with constant stirring, in 100 g paraffin heated to smoking.
Embedding wax
Paraffin	100 g
Stock rubber solution	4 g
Beeswax	1 g

Maxwell's Embedding Wax:

Paraffin	100 g
Hance's stock rubber solution	4 g
Bayberry wax	7 g
Beeswax	1 g

The Technique of Dehydrating, Clearing, and Embedding. Before passing to the choice of a microtome and the method of using it, it is necessary to discuss briefly the actual operations involved in using the dehydrating, clearing, and embedding media selected. The techniques of dehydration and dealcoholization do not differ materially from those used in the preparation of wholemounts, which have been described. The whole process, however, could be much simplified if people would only remember that water is heavier than the majority of dehydrating agents and that the majority of dehydrating agents are lighter than most clearing agents. In translating this theory into practice, it must be obvious that the object to be dehydrated should be suspended toward the top of a tall cylinder of dehydrant, in order that the water extracted from it may fall toward the bottom of the vessel, and that an object to be cleared should be held at the bottom of the vessel for the reverse reason. It is, indeed, practically impossible to dehydrate a large object unless it is so suspended. The process of impregnating the tissues with wax has not been discussed previously and will be dealt with fully.

The first prerequisite for embedding is some device that will maintain the temperature of the wax at just above its melting point. Most people employ complex thermostatically controlled ovens for this purpose, but the simple device shown in Fig. 12-5 is practical and cheap. As will be seen, this device consists essentially of a series of incandescent electric bulbs held at a distance, which may be varied, above a series of glass vials. Before the embedding process is started, as many vials as will be required are filled with wax and placed under the reflector, and the current is turned on. After a little while, it will be observed that the absorbed heat has melted the wax. The wax may be melted only at a small surface layer, throughout the entire vial, or, as is required, in the upper two-thirds of the vial. If this last is not achieved, the height of the lamp must be varied until after an hour or two each of the vials contains about one-third unmolten opaque wax at the bottom and two-thirds clear molten material above. Thus, when the object is placed in one of these vials, it will drop until it reaches the solidified layer, where it will remain in contact with molten wax at exactly the melting point of the wax. It is obvious that the room in which this operation is to be conducted must be at a fairly constant temperature and must be relatively free of drafts, but only a very large volume of embedding work justifies the purchase of an expensive thermostatically controlled oven. If such an oven is to be purchased, it is highly desirable to avoid one in which the heat is distributed by convection. The oven shown in Fig. 12-6, in which a circulating fan continuously moves the air and thus maintains a uniform temperature throughout the whole oven, is infinitely to be preferred. It is the high cost of such circulating-air ovens that leads

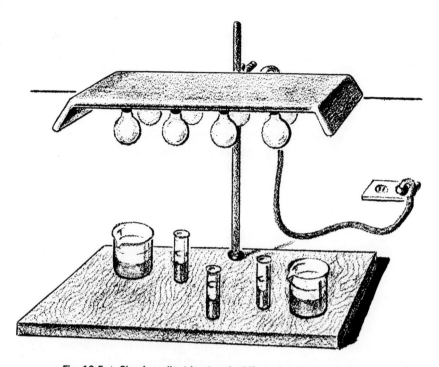

Fig. 12-5 | Simple radiant-heat embedding oven. Height of the hood should be adjusted until the wax is melted for about one-half its depth.

the author to believe that much more use should be made of the simple radiant-heat embedding device just discussed.

Assuming that the material has been passed through dehydrating and clearing agents and is now awaiting embedding, there are two main methods by which this may be done. Either the object may be transferred directly to a bath of molten wax or it may be passed through a graded series of wax-solvent mixtures. The author is strongly in favor of the latter course. Let us suppose that we have selected benzene as the clearing agent and that the object is in a vial containing a few milliliters of this solvent. Take a block of whatever medium is to be used for embedding and shave a few chips from it with a knife. Add these to the solvent. The chips usually dissolve very slowly and form a thickened layer at the bottom of the tube through which the object to be embedded sinks. The average object will be satisfactory if left overnight. Then place the tube itself in the embedding oven maintained at a temperature somewhat above that of the melting point of the wax and add as many more shavings as can be crammed into the tube. When these are com-

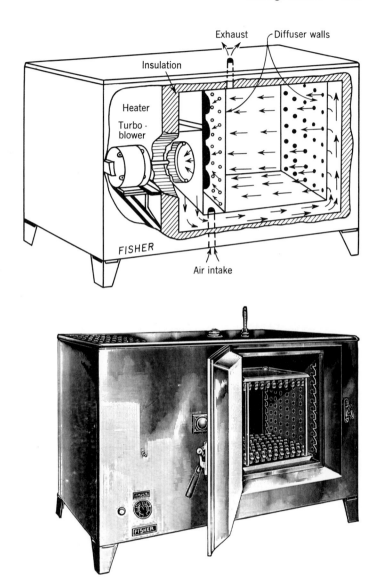

Fig. 12-6 | Circulating-air embedding oven.

pletely melted and a large quantity of the volatile solvent has evaporated, remove the object with a pipette or forceps and place it in a dish of pure wax for an hour or two before transferring it to a second dish of pure wax for the time necessary to secure complete impregnation.

There is no method of forecasting how long an object will take to become completely impregnated with wax. It is very easy to find out (see Table 2) when one has started to cut sections whether or not the impregnation is complete, but there is no basis save experience on which to establish the timing in the different baths of wax. If the object is to be transferred directly from solvent to wax, at least three baths should be employed, since nothing is more destructive to a good section than the presence of a small quantity of the clearing agent in the embedding medium. To an absolute beginner seeking a rough guess, it may be said that a block of liver tissue, 3 to 5 mm in size, will be satisfactorily impregnated with wax after 30 min in each of three baths, whereas a 96-hr chick embryo will require at least 2 hr in each of three baths for its successful impregnation.

While the object is being impregnated with the wax, it is necessary to decide what type of vessel will be used to cast the final block. This will depend to a far greater extent on the size of the object than on the preference of the worker. Very small objects may be embedded most satisfactorily in ordinary watch glasses (that is to say, ordinary thin-walled watch glasses, not syracuse watch glasses of the laboratory type) or in any other thin-walled glass vessel. Very large objects are often embedded with the aid of two thick L-shaped pieces of metal which, when fitted together, form a rectangular mold of varying dimensions. The author regards these as very clumsy and would always prefer to prepare a cardboard or paper box rather than to endeavor to maneuver metal molds which are always getting jarred out of place at the wrong moment. The preparation of a paper box is easy; the method preferred by the author is shown in Figs. 12-7 to 12-13.

After the box has been prepared, the actual process of embedding is begun. This is shown in great detail in Figs. 12-14 to 12-17. Before starting, it is necessary to provide the following items: (1) a dish of water of sufficient size that the finished block may be immersed in it readily; in the illustration an ordinary laboratory finger bowl is in use, (2)

Fig. 12-7 | Laying out paper for an embedding box.

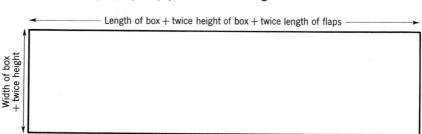

Making Sections 157

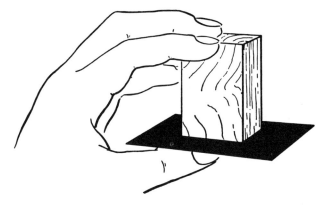

Fig. 12-8 | **Folding a paper box.** The block is centered on the sheet cut as in Fig. 12-7.

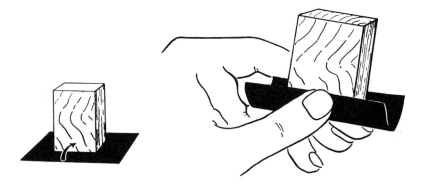

Fig. 12-9 | **Folding a paper box.** The sides are folded up.

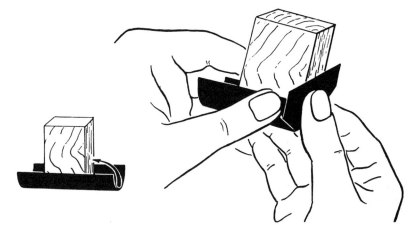

Fig. 12-10 | **Folding a paper box.** The end is folded up.

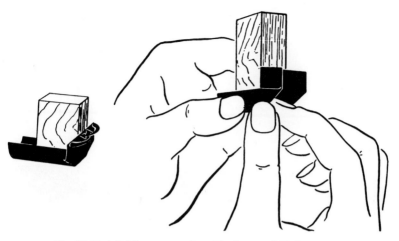

Fig. 12-11 | **Folding a paper box.** The flaps are folded in.

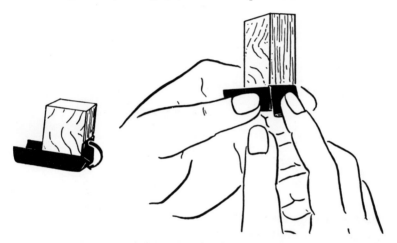

Fig. 12-12 | **Folding a paper box.** The end is folded down and creased.

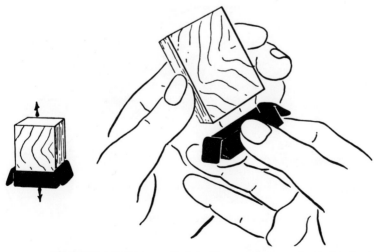

Fig. 12-13 | **Folding a paper box.** The cycle is repeated with the other end and the finished box removed.

Fig. 12-14 | Filling with wax an embedding box which has been attached with water to a glass slide.

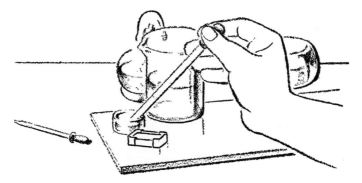

Fig. 12-15 | Transferring the object from the embedding dish to the wax-filled paper box.

Fig. 12-16 | Remelting the wax around the object with a heated pipette.

Fig. 12-17 | Cooling the wax block.

some form of heat, an alcohol lamp being just as effective as a bunsen burner, (3) a large slab of plate glass, and (4) a wide-mouthed eyedropper type of pipette. The oven employed should contain the object itself in a small container of molten wax, as well as another container of the medium used. It must be emphasized that one cannot impregnate an object with one kind of wax and embed it in another. The first thing to do is to wet the *underside* of the bottom of the paper box in the water and press it firmly into contact with the plate-glass slab, which will both hold it in position and assist in cooling the wax. Then take (Fig. 12-14) one of the beakers of molten embedding material from the oven and fill the little paper box to the brim. Heat the pipette in the flame to a temperature well above that at which the wax will melt. Use it to pick up the object from its own dish (Fig. 12-15) and to transfer it to the paper box. By the time that this has been done, a layer of hardened wax will have been formed at the bottom of the paper box, so that the object will rest on the layer of solidified wax with a molten layer above it. It will happen almost invariably that the surface has also cooled, so that a crust of cool wax will have been carried down with the object into the box. It is essential to get rid of this if the wax is to adhere through section cutting. Reheat the pipette; use it to melt the entire surface of the box (Fig. 12-16) and to maneuver the object into the approximate position in which it is required to lie in the finished block. Then blow on the surface until the wax is sufficiently solidified to enable you to pick up the box carefully and, as shown in Fig. 12-17, hold it on the surface of the water used for cooling. With most wax media it is desirable to cool the block as rapidly as possible, and it should never be permitted to cool in air. It cannot, however, be pushed under the surface of the water because the molten center is liable to break through the surface crust and thus destroy the block. After holding the box in the position indicated until it is fairly firm throughout, push it under the surface to complete the cooling. Large blocks should be kept under the surface with weights.

The block may be left in water for any reasonable length of time, but if it is to be stored for days or weeks it is much better kept in a 5 per cent solution of glycerin in 70 per cent alcohol. There seems to be a widespread delusion that because an object must be perfectly dehydrated before being impregnated with wax, it must subsequently be kept out of contact with fluids. Nothing could be further from the truth. As will be discussed later, when dealing with the actual technique of sectioning, it is often desirable to expose a portion of the object to be sectioned and leave it under the surface of water for some days in order to get rid of the brittleness that has been imparted through the embedding process. Blocks that have been stored dry for a long period of time should always be soaked in a glycerin-alcohol mixture for at least a day before sectioning. This is particularly important with plant material.

It is, in any case, undesirable to section a block as soon as it has been made because it is necessary for successful sectioning that the block be the same temperature throughout. If a block is made in the evening, it is better to take it out of the water and to leave it lying on the bench overnight so that the temperature may be stabilized.

Choice of a Microtome. Microtomes may be broadly divided into two classes. In the first of these the block remains stationary, while the knife is moved past it. In the second group are those in which the block moves past a stationary knife. The first class (Fig. 12-18) is commonly known as a *Schantz microtome* (after the original German model). They are made by several manufacturers and are rarely used for the preparation of serial sections. They have the advantage that relatively large blocks may be cut, but they have the disadvantage that no ribbon can be obtained broader than the width of the knife.

A Spencer or rotary microtome is shown in Fig. 12-19. In this type of microtome, the rotation of the large wheel causes the block holder to move vertically up and down in most instances through a distance of about 3 in. The portion that slides up and down has, at the end opposite the block, a rectangular plate of hardened steel inclined at an angle

Fig. 12-18 | Sliding microtome.

Fig. 12-19 | American Optical Company rotary microtome.

of about 45°. Under the pressure of a powerful spring, this plate bears against a hardened steel knob, which is itself connected to a micrometer screw. As the handle is rotated, a pawl works against a ratchet to move the micrometer screw and thus the knob connected with it through a given distance for each rotation. As the knob moves forward, it bears on the diagonal plate, which moves the block the required distance forward at each revolution. This mechanism is very costly to make and is liable to have a large number of minor defects that are not always apparent until one has started section cutting. One of the most important things to watch is that the knob which controls the section thickness is so moved that an exact number of microns is indicated. If, for example, the knob is moved so that the indicator line lies between 9 and 10 μ, the pawl will not engage the ratchet perfectly but will chip off a small portion of brass at each revolution. It requires only a few weeks of operation under these careless conditions to destroy the ratchet wheel, which will have to be replaced at the factory. No inexperienced student should ever be trusted with one of these machines until its mechanism has been explained and clearly demonstrated to him.

Knives and Knife Sharpening. The most important single factor in the production of good sections is the knife used in cutting. It does not matter how much care has been taken in the preparation of the block or how complex a microtome is used; if the knife-edge is not perfect, there is no chance of obtaining a perfect section. Ordinary razors are not satisfactory for the production of fine sections. It is necessary to buy a microtome knife, preferably from the manufacturer of the microtome. Another type of microtome knife employs the edge of a safety-razor blade in a special holder; this does not, in the author's opinion, give so good results as a solid blade.

Three types of solid blades are available. First are those that are square-ground, that is, those in which the main portion of the knife is a straight wedge. Second are those that are hollow-ground, that is, those in which both sides of the knife have been ground away to a concave surface, resulting in a relatively long region of thin metal toward the edge. Third are knives that are half-ground, that is, knives in which one side is square- or flat-ground and the other side is hollow-ground. This last type of knife, which the author prefers, is a compromise. There is no doubt that a square-ground knife is sturdier than a hollow-ground knife, a point of some importance when cutting large areas of relatively hard tissues, but there is also no doubt that a hollow-ground knife can be brought more readily to a fine edge. Where a half-ground knife is employed, the flat side should always be toward the block. Microtome knives must be sharpened frequently, but it is necessary before discussing how to do this to give a clear account of the nature of the cutting edge itself.

If a wedge of hardened steel were to be ground continually to a fine edge, as in Fig. 12-20, it would be utterly worthless for cutting. After only a few strokes, the fine feather edge, which would be produced by this type of grinding, would break down into a series of jagged sawteeth. A microtome knife, or, for that matter, any other cutting tool, must have ground on its cutting edge a facet of a relatively obtuse angle, whether it is a square-ground knife or a hollow-ground knife (both are shown in

Figs. 12-20 and 12-21 | **Types of cutting edge.** Fig. 12-20. Simple wedge without cutting facet. Fig. 12-21. Flat- and hollow-ground blades showing cutting facet.

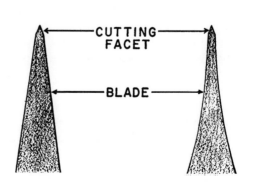

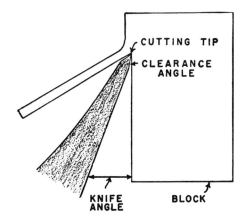

Fig. 12-22 | Diagram showing cutting action of knife on wax block.

Fig. 12-21). The process of applying this cutting facet to the tip is known as *setting*. It is an exceedingly difficult operation to conduct but one that must be learned by every user of a microtome knife. The actual grinding of the blade itself to the correct angle or to the correct degree of hollowness cannot be done in a laboratory; the knife must be returned to the manufacturer or to some scientific supply house equipped with the special machinery necessary. The cutting facet, however, must be set at least once a day if the blade is in continual use. The nature and purpose of this cutting facet are best explained by reference to the technique of cutting shown in Fig. 12-22. Notice first that the knife blade itself must be inclined at such an angle to the block that the cutting facet is not quite parallel to the face of the block. There must be left a clearance angle to prevent the knife from scraping the surface every time that it removes a section. In cutting wax, this clearance angle should be as small as possible, and it is for this reason that the blade holder of a microtome is furnished with a device for setting the knife angle. The knife angle should not be set with reference to any theoretical consideration but with regard only to securing this small clearance angle. The only way to judge whether or not a satisfactory clearance angle has been obtained is to observe the sections as they come from the knife. If the clearance angle is too large, so that the section is not being cut from the block but is being scraped from it, the section will have a wrinkled appearance and will usually roll up into a small cylinder. If the clearance angle is too small, so that the lower angle of the facet is scraping the block after the tip is passed, the whole ribbon of sections will be picked up on the top of the block, which will itself crack off when the knife point reaches it. It is obvious that the knife angle will be changed as the angle of the cutting facet is changed, so that it is desirable to maintain

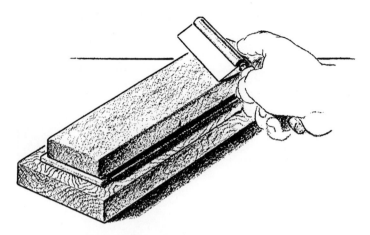

Fig. 12-23 | **Setting the cutting facet.** The razor is drawn gently forward.

the cutting facet at as uniform an angle as possible. This angle is set on the knife in the manner shown in Fig. 12-23. Notice that the knife has been furnished with a handle and that a small split cylinder of steel has been slipped over the back of the blade. This split cylinder rests flat on the stone, as does the edge of the blade, so that when the knife is pushed forward (Fig. 12-23 shows it at the beginning of the stroke) the cutting facet is produced as the angle between the cutting edge lying on the stone and the enlarged temporary back that has been placed on the knife. Since a much blunter cutting facet is required for hard materials than for soft, it is strongly recommended that either two knives or at least two sharpening backs be obtained. It does not matter what kind of stone is used for sharpening, provided that it is of the finest obtainable grit, that it is absolutely flat, and that under no circumstances is it used for any purpose except the sharpening of microtome knives. It does not matter whether it is a waterstone, lubricated with glycerin, like the Water-of-Ayr stones generally used in Europe, or an oilstone, lubricated with mineral oil, like the Pike stones so commonly employed in the United States. However, it is important that it be flooded with lubricant before starting and that the knife be drawn with a light pressure (notice that the finger is *behind* and not on top of the knife in the illustration) the entire length of the stone at each operation. If only the central portion of the stone is used, it soon becomes hollowed out, and it is thus impossible to maintain a uniform angle. About three strokes on each side of the knife are quite enough to produce a perfectly sharp cutting facet, and further strokes will have no effect other than to diminish the length of life of the knife.

This direction for the use of three strokes in setting applies, of course, only to a knife that has been treated reasonably and not to one which through carelessness has acquired a nick in its edge. Where the nick is large, it is almost impossible to remove it in setting because the continual setting merely grinds away the edge of the knife and ultimately alters the thickness of the blade itself. If the knife-edge is nicked to a deeper extent than about ¼ mm, the only thing to do is either to return the knife to the manufacturer to be reground or to avoid that portion of the blade containing the nick when cutting sections. It must be emphasized that the only purpose of setting is to produce a cutting facet and that grinding, which cannot be done in the ordinary laboratory, is required for the removal of knife imperfections. The next question is that of stropping the blade of the knife by pulling it backward across a leather surface in the manner shown in Fig. 12-24. If the knife has been set properly, stropping, the only purpose of which is to polish the facet, is quite unnecessary. The nature of the leather surface that is used for stropping makes it obviously impossible to pull the knife blade forward, and there is a grave risk in pulling it backward that the facet, instead of becoming polished on its flat surfaces, will become rounded on its edges and thus undo the work of setting. Certainly no beginner should be permitted to use a strop until he has demonstrated his ability to set a knife-edge to the point where it will cut an excellent section without stropping. It is also strongly recommended to the beginner that he examine the edge of a knife under the low power of a microscope before setting, after setting, and after stropping.

Fig. 12-24 | **Stropping a microtome knife.** The knife is drawn smoothly backward once or twice on each side.

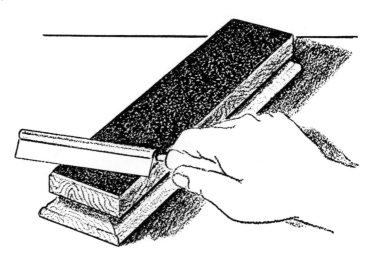

Several machines for sharpening knives are available, but their cost is so high that their purchase can be justified only in the largest laboratories. Moreover, many of them require nearly as much skill as hand sharpening. Many companies offer a sharpening service, which is convenient when several knives are available.

Mounting the Block. After the knife has been sharpened and the microtome has been selected, the block is trimmed to the correct shape and attached to the object holder of the microtome. The rough block of wax containing the object first must be removed from whatever was used to cast it in or, if a paper box was used, the box cut away roughly with a knife. The block should now be held against a light so that the outlines of the contained object can be seen clearly. The block is trimmed until the object lies in the center of a perfect rectangle with the major axis of the object exactly parallel to the long sides. This is best achieved by finding the major axis at right angles to which the sections are to be cut and by trimming down one side of the block with a sharp safety-razor blade, taking off only a little wax at a time. If one tries to remove a large quantity of wax, there is the danger of cracking the block. After one side is shaved to a flat surface, the other side is shaved parallel to it. The top and bottom surfaces of the block may now be shaved, and it is essential that these be exactly parallel to each other. A skilled microtomist can cut these edges parallel with a safety-razor blade without very much difficulty, but numerous devices have been described from time to time in the literature to enable one to do this mechanically. It does not matter if these two edges are exactly parallel to the plane of the object; it is essential only that they be parallel to each other. At this stage plenty of wax should be left both in front of and behind the object.

This trimmed block now has to be attached to a holder that can be inserted into the microtome. Since the majority of sections today are cut on a Spencer rotary microtome, the following description is of the use of one of the holders supplied with this machine, although the ingenuity of man has not yet succeeded in devising a worse method of attaching a paraffin block to a microtome. The holder, which is seen in Fig. 12-25, consists of a disk of metal with a roughened surface attached to a cylindrical shank. First of all, this disk must be covered with a layer of wax, and it is extraordinarily difficult to get wax to adhere to these chromium-plated surfaces. If the worker is not entirely bound by convention, it would be much better for him to obtain a series of small rectangular blocks of some hardwood like maple and to soak these for a day or two in molten wax. After they are removed, drained, and cooled, it is the simplest thing in the world to attach a paraffin block to them and to hold them in the jaws of the microtome. Whether the metal holder or the wooden one is used, the technique is essentially the same. A layer of

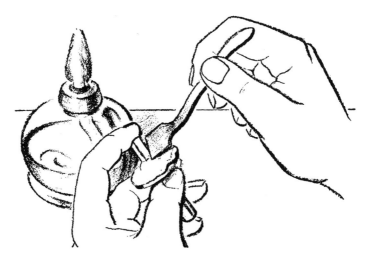

Fig. 12-25 | Mounting the wax block on the block holder.

molten wax is built up on the surface and allowed to cool. The block (see Fig. 12-25) is pressed lightly onto this hardened wax and fused to it with the aid of a piece of heated metal. Some people use old scalpels, but the author prefers the homemade brass tool shown in Fig. 12.25. Care must be taken to press very lightly with the forefinger and to conduct the whole operation as speedily as possible to avoid softening the wax in which the object is embedded. The metal tool should be heated to a relatively high temperature and applied by just touching it lightly. If the block is very long, it is also desirable to build up small buttresses of wax against each side, being careful not to bring these buttresses so far up the block that they reach the tip of the object to be cut. The metal should now be put aside and allowed to reach room temperature. Many people at this point throw the block and holder into a finger bowl of water, which is all right provided the water is at room temperature. However, there is no more fruitful source of trouble in cutting sections than to have the knife, the block, and the microtome at different temperatures. It is much better to mount the blocks the day before they are to be cut and to leave them on the bench to await treament. Then final inspection is made of the block to make certain that the upper and lower surfaces of the block are flat, smooth, and parallel. Many people do not make the final cuts on these surfaces until after the block has been mounted in the block holder. The block and the block holder, after insertion in the jaws of the microtome, are seen in Fig. 12-26. It will be noticed that set screws on the apparatus permit universal motion to be imparted to the block, so that it can be orientated correctly in relation

Fig. 12-26 | Starting the paraffin ribbon.

to the knife. It is easy to discover whether or not the edges are parallel by lowering the block until it does not quite touch the edge of the knife, adjusting it until the lower edge is parallel, lowering the block again, and then comparing the relation of the upper edge with the edge of the razor.

Cutting Paraffin Ribbons. The first step in cutting sections on this type of microtome is to make sure that every one of the set screws seen in Fig. 12-26 is completely tight. The set screws holding the block holder may be tightened in any order, provided that the result leaves the block correctly oriented, but those connected with the knife must be done in the correct order. First the knife is inserted into the holder and fixed firmly but not tightly in place by the two bearings at each end. The tightening of these screws causes the two movable holding arms to hold the knife near its edge. The knife is held in a pair of hemicylinders, which may be moved to adjust the knife angle (see Fig. 12-22). The knife should be set at the angle that experience has shown to be desirable —no guide other than experience can be used—and the two set screws that lock these inclinable hemicylinders in place then tightened. While the knife is held in place, the two original set screws should be screwed as tightly as the thumb can bear. This leaves the two set screws, which come through the inclinable hemicylinders and bear on the back of the knife. These two set screws should be tightened simultaneously and uniformly. The effect of this is to force the knife upward and thus wedge it with extreme firmness in the knife holder.

After all is tight, the handle on the microtome is turned until the block is as far back as possible, and the entire knife moved on its carriage until the edge of the blade is about ¼ mm in front of the block. A last-minute check is now made to make sure that the divisions of the setting device exactly coincide with the thickness desired. The handle is rotated rapidly until the block starts cutting. The front face will rarely be parallel to the blade of the knife, so that a considerable number of sections will have to be cut until the entire width of the block is coming against the knife. No particular attention need be paid to the quality of this initial ribbon, which may be thrown away.

If all is not going well and the ribbon is not coming off in a perfect condition, refer to Table 2. The remaining operations of preparing and mounting the ribbon are seen far more clearly in illustration than by description. As soon as the ribbon is the width of the knife in length, a dry soft brush, held in the left hand, is slipped under the ribbon, which is then raised in the manner shown in Fig. 12-26. Care should be taken that a few sections always remain in contact with the blade of the knife because if the ribbon is lifted till only the edge of the section lies on the edge of the knife, the ribbon will break almost invariably. As the handle is turned, the brush in the left hand is moved away until the ribbon is the length of whatever sheet of paper one has to receive it on. Legal-size (foolscap) paper is employed quite commonly and is shown in Fig. 12-27. Notice that the left-hand edge of the ribbon has been laid flat some distance from the edge of the paper and that a loop large enough to avoid strain on the ribbon attached to the knife is retained with the brush, while the ribbon is cut with a rocking motion of an ordinary scalpel or cartilage knife. The larger and colder this scalpel is, the less likelihood there is of the section adhering to it. The purpose of leaving a good margin around the edge of the paper is that it may be desirable to interrupt ribbon cutting for a time and to continue later. In this case the worker should furnish himself with a little glass-topped

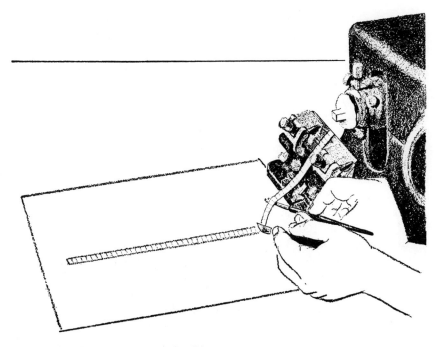

Fig. 12-27 | Laying out the ribbon.

frame, which is laid over the paper to prevent the sections from being blown about. As the inexperienced worker will soon find out, the least draft of air, particularly the explosive draft occasioned by someone opening the door, is sufficient to scatter the ribbons all over the room. These operations of carrying the ribbon out with the left hand, transferring the brush to the right hand, and cutting the ribbon off are continued until the whole of the required portion of the block has been cut and lies on the paper.

The ribbon must be divided into suitable lengths for mounting on a slide (Fig. 12-28). Although in theory a section should be of the same size as the block from which it came, this practically never occurs in practice, and it is usually safe to allow at least 10 and sometimes 20 per cent for expansion when the sections are finally flattened. The ribbon should never be cut completely until a sample has been flattened on a slide, in order that one may judge the degree of expansion. Although the sections shown in Fig. 12-29 are mounted on an ordinary 3- by 1-in. slide, it would be more practical for a ribbon as wide as this to use a 3- by 1½-in. or even a 3- by 2-in. slide. The sections should never occupy the whole area of the slide. At least ¼ in. should be left at one end for subsequent labeling. When the decision has been made as to how many sections shall be left in each piece of ribbon, the first row of ribbons is cut into the required lengths (Fig. 12-28). Then the worker must decide what shall be used to cause them to adhere to the slide. It is conventional to use:

Mayer's Albumen:

Fresh egg white	50 ml
Glycerin	50 ml
Sodium salicylate	1 g

The author prefers to dilute the selected adhesive two or three hundredfold with water and to use this diluted adhesive in the next operation of flattening the sections. However, the adhesive can be made full strength and used immediately. If this procedure is desired, shake the ingredients together until they are thoroughly mixed. Filter. Apply a thin smear of this on a clean slide with the tip of the little finger.

Fig. 12-28 | Cutting the ribbon in lengths.

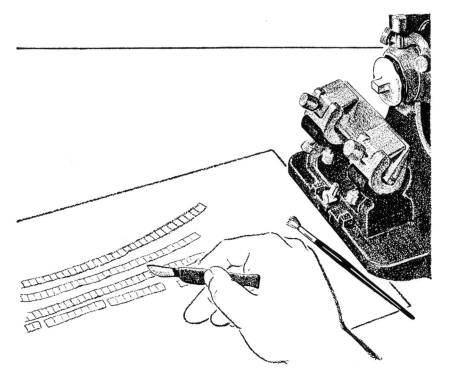

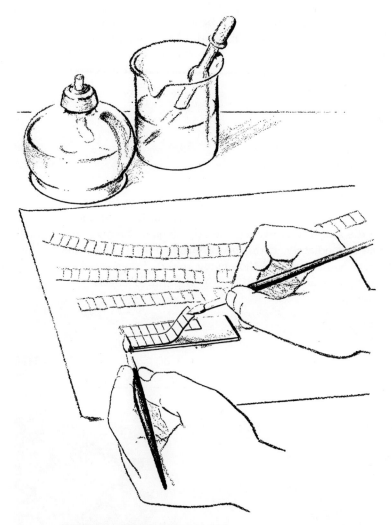

Fig. 12-29 | Mounting the dry ribbon.

Many people prefer to mount large sections with a gelatin adhesive in place of the albumen. One of the best of these is:

Weaver's Gelatin Solution:

Solution A
    Water   10 ml
    Calcium propionate   0.1 g
    Roccal   0.1 ml
    Gelatin   0.1 g

Solution B
　Water　　　　　　　　90 ml
　40% formaldehyde　　10 ml
　Chrome alum　　　　　1 g

Roccal is a proprietary bacteriacide manufactured by the Rohm and Haas Company, and calcium propionate, also sold under many trade names, is a fungus inhibitor widely used in the baking trade. The two solutions are mixed, immediately before use, in the proportion 9 parts solution B to 1 part solution A. Sections are spread and flattened on the mixture just as they are when diluted egg albumen is used.

It will have been apparent to the worker from the moment he started cutting the sections that they are not absolutely flat. They may be slightly crinkled or slightly distorted and, therefore, must be flattened by being warmed on water heated just below the melting point of the wax. Some people place this water on the slide and then add the sections to it, but the author prefers to lay the ribbons on the slide as shown in Fig. 12-29. This is not nearly so easy as it looks. The tips of two brushes should be moistened with the tongue just enough to bring the hairs to a point. Then the two moist points are delicately touched (too much pressure will cause the ribbon to adhere to the paper) to each end of the selected piece of section. This piece is lifted as shown in the illustration and placed on the slide. When a sufficient number of pieces of ribbon have been accumulated, the slide is picked up carefully, reversed, and laid on top of the last three fingers of the left hand exactly as shown in Fig. 12-30. It is fatal to grasp the slide by the sides; if this is done when the water is flooded on from the pipette, the meniscus coming to the edge of the slides will break against the fingers, causing the sections to adhere to the fingers permanently. The manner shown is quite safe. The water containing the adhesive (if none has been applied to the slide) is flooded on from a pipette as illustrated. Plenty of fluid should be applied and should be raised in quite a sharp meniscus from the edge of the slide.

The sections must now be flattened. This is much better done rapidly with a flame than slowly on a hot plate. Figure 12-31 shows the slide held over an alcohol lamp, but a microbunsen can be employed equally well. The slide should be exposed to heat for a moment, withdrawn to give time for the heat to pass from the glass to the fluid, rewarmed, and so on until the sections are observed to be flat. The utmost care must be taken at this point, for if the paraffin is permitted to melt, it will be difficult later, if not impossible, to cause the sections to remain attached to the glass. As soon as the sections are flattened, the slide is gently tilted backward toward the hand to run off the excess water against the thumb, leaving the sections stranded in place. The slide is usually placed on a thermostatically controlled hot plate (seen in Fig. 12-39) and permitted

Fig. 12-30 | Flooding the ribbons.

to dry. Most people leave their slides overnight, but frequently an hour would be sufficient. Dryness can be gauged without the least trouble by the fact that a moistened slide shows the wax to be more or less opalescent, while on a properly dried slide it is almost glass-clear.

The method just described is susceptible of several variations, which may be noted briefly. Some people neither drain the water from the slide nor heat the slide over the lamp but merely place it, as soon as the water has been added to it, on the thermostatically controlled hot plate. This permits the sections to dry and to flatten at the same time. The objection to this procedure is that air contained in the water used for flattening almost invariably comes out in the form of bubbles, which accumulate under the sections, either causing them to fall off or at least making it very difficult to observe them properly when mounted. There is also the risk in this procedure that the water will not stop at the edge of the slide but will flood off it unexpectedly, carrying the sections with it onto the surface of the hot plate.

Another procedure, which can be recommended for all slides but

Fig. 12-31 | Warming the flooded ribbons to flatten them.

which is essential for those flattened on chrome gelatin, is to blot the sections before putting them on the hot plate. After the excess fluid has been drained from the slide, the latter is laid on a smooth surface, and a sheet of *water-saturated coarse* filter paper is laid firmly on top. A soft rubber roller, of the type used to attach photographs to glazing plates, is now run backward and forward over the slide with a firm pressure. This assures that the sections will be perfectly flattened in contact with the slide but requires strong nerves to try for the first time because of the fear that the sections will stick to the paper. This has not happened in a good many thousand slides which the author has made by this means, and slides so prepared are always free from air bubbles.

Before proceeding to a discussion of the next steps to be taken, it may be well to review the innumerable things that may happen to prevent the production of a perfect ribbon. The appearance, cause, and cure of the more common defects are shown in Table 2. These are by no means the only defects or the only cures that may be applied, and every user of the microtome should have in his hands O. W. Richards's "The Effective Use and Proper Care of the Microtome," obtainable from the American Optical Company, which lists many suggestions beyond those here given.

TABLE 2 | Defects Appearing while Sections Are Being Cut

Defect	Possible causes	Remedies
 Fig. 12-32 | Ribbon curved.	1. Edges of block not parallel. 2. Knife not uniformly sharp, causing more compression on one side of block than other. 3. One side of block warmer than other.	1. Trim block. 2. Try another portion of knife-edge or resharpen knife. 3. Allow block to cool. Check possible causes of heating or cooling, such as lamps or drafts.
 Fig. 12-33 | Sections compressed.	1. Knife blunt. 2. Wax too soft at room temperature for sections of thickness required. 3. Wax warmer than room temperature.	1. Try another portion of knife-edge or resharpen knife. Compression often occurs through a rounded cutting facet (see Fig. 12-21) produced by overstropping. 2. Reembed in suitable wax or cut thicker sections. Cooling block is rarely successful. 3. Cool block to room temperature.
Fig. 12-34 | Sections alternately thick and thin usually with compression of thin sections.	1. Block or wax holding block to holder still warm from mounting. 2. Block or wax holding block to holder cracked or loose. 3. Knife loose. 4. Knife cracked. 5. Microtome faulty.	1. Cool block and holder to room temperature. 2. Check all holding screws. Remove block from holder and holder from microtome. Melt wax off holder and make sure holder is dry. Recoat holder and remount block. Cool to room temperature. 3. Release all holding screws and check for dirt, grit, or soft wax. Check knife carriage for wax chips on bearing. 4. Throw knife away. 5. Return microtome to maker for overhauling.

TABLE 2 (Continued) | Defects Appearing while Sections Are Being Cut

Defect	Possible causes	Remedies
 Fig. 12-35 | Sections bulge in middle.	1. Wax cool in center, warm on outside. 2. Only sharp portion of knife is that which cuts center of block. 3. Object impregnated with hard wax and embedded in soft, or some clearing agent remains in object.	1. Allow block to adjust to room temperature. This is the frequent result of cooling blocks in ice water. 2. Try another portion of knife-edge or resharpen knife. 3. Reembed object.
 Fig. 12-36 | Object breaks away from, or is shattered by, knife.	1. If object appears chalky and shatters under knife blade, it is not impregnated. 2. If object shatters under knife but is not chalky, it is too hard for wax sectioning. 3. If object pulls away from wax but does not shatter, the wrong dehydrant, clearing agent, or wax has been used.	1. Throw block away and start again. If object is irreplaceable, try dissolving wax, redehydrating, reclearing, and reembedding. 2. Spray section between each cut with celloidin. 3. Reembed in suitable medium, preferably a wax-rubber-resin mixture. Avoid xylene in clearing muscular structures.
Fig. 12-37 | Ribbon splits.	1. Nick in blade of knife. 2. Grit in object.	1. Try another portion of knife-edge. 2. Examine cut edge of block. If face is grooved to top, grit has probably been pushed out. Try another portion of knife-edge. If grit still in place, dissect out with needles. If much grit, throw block away.

TABLE 2 (Continued) | Defects Appearing while Sections Are Being Cut

Defect	Possible causes	Remedies
Fig. 12-38 | Block lifts ribbon.	1. Ribbon electrified. (Check by testing whether or not ribbon sticks to everything else.)	1. Increase room humidity. Ionize air with either high-frequency discharge or bunsen flame a short distance from knife.
	2. No clearance angle (see Fig. 12-22).	2. Alter knife angle to give clearance angle.
	3. Upper edge of block has fragments of wax on it (a common result of 2).	3. Scrape upper surface of block with safety-razor blade.
	4. Edge of knife (either front or back) has fragments of wax on it.	4. Clean knife with xylene.

No ribbon forms

Defect
1. Wax crumbles.
2. Sections, though individually perfect, do not adhere.
3. Sections roll into cylinders.

Possible causes
1. Wax contaminated with clearing agent.
2. Very hard pure paraffin used for embedding.
3a. Wax too hard at room temperature for sections of thickness required.
 b. Knife angle wrong.

Remedies
1. Reembed. (*Note:* Wax very readily absorbs hydrocarbon vapors.)
2. Dip block in soft wax or wax-rubber medium. Trim off sides before cutting.
3a. Reembed in suitable wax. If the section is cut very slowly and the edge of the section held flat with a brush, ribbons may sometimes be formed.
 b. Adjust knife angle.

Staining and Mounting Sections. Assuming that all the difficulties mentioned in the last section have been overcome and that one now has a series of slides bearing dried consecutive ribbons, the next thing to be done is to remove the paraffin in order that they may be stained. It is conventional, although probably not necessary, to warm each slide over a flame (holding it as shown in Fig. 12-31) until the paraffin is melted thoroughly. Then the slide is dropped, as shown in Fig. 12-39, into a jar containing xylene, benzene, or some other suitable paraffin solvent.

It is necessary through the subsequent proceedings to be able to recognize instantly that side of the slide on which the section lies. This is not nearly so easy as it sounds, and a lot of good slides have been lost by having the sections rubbed off them. The simplest thing to do is to in-

Fig. 12-39 | Starting a slide through the reagent series.

cline the slide at such an angle to the light that, if the section is on top, a reflection of the section is seen on the lower side of the slide. A diamond scratch placed in the corner is of little use because it becomes invisible when the slide is in xylene. The greatest care should be taken to remove all the wax from the slide before proceeding further, and it is usually a wise precaution to have two successive jars of xylene, passing the second jar to the position of the first and replacing it with fresh xylene after about ten or twelve slides have passed through. It must be remembered that paraffin is completely insoluble in the alcohol used to remove the xylene, so that it is of no use to soak a slide in a solution of wax in xylene and imagine that it will be sufficiently free from wax for subsequent staining. Some people use three jars, the first two containing xylene and the third having a mixture of equal parts of absolute alcohol and xylene, to make sure that all the wax is removed. If even a small trace of wax remains, it will prevent the penetration of stains. Assuming that one is proceeding along the classic xylene-alcohol series, the slide is then passed from either the fresh xylene or the xylene–absolute alcohol mixture to a coplin jar of absolute alcohol. It is unfortunate that as yet nobody seems to have placed on the market a coplin jar or slide-staining dish that has a lid which fits tightly, since absolute alcohol, which is very hygroscopic, is rarely of much use after it has been left on an open bench for a day or two. It does not matter much if xylene is carried over into the absolute alcohol, but as soon as the first trace of a white flocculent precipitate appears in the alcohol—indicating that some wax is being carried over—it must be replaced by fresh alcohol.

The author never bothers to use a series of graded alcohols between absolute alcohol and water. These graded series are necessary, of course, when one is dealing with the dehydration of whole objects that may be distorted, but the author has never been able to find the slightest difference between a thin section that has been passed from absolute alcohol to water and one that has been graded laboriously down through 90 per cent, 80 per cent, etc., the length of the series varying with the wishes of the individual. As soon as the slide has been in water long enough to remove the alcohol, it should be withdrawn and examined carefully to make sure that it has been sufficiently dewaxed. If the water flows freely over the whole surface, including the sections, it is safe to proceed to staining in whatever manner is desired. If, however, the sections appear to repel the water or if there is even a meniscus formed around the edge of a section, it is an indication that the wax has not been removed and that the slide must again be dehydrated in absolute alcohol, passed back into a xylene-alcohol mixture, and again put into pure xylene.

In the specific examples in Part Three, descriptions are given of individual staining methods. The purpose of this chapter is to discuss only the general principles involved.

Fig. 12-40 | Placing the coverslip on serial-section slide.

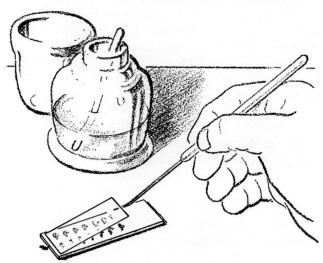

It is assumed that the sections will be mounted in a solution of dried balsam in xylene. The slide is removed from the xylene, drained, and placed on any convenient flat surface. A drop of the mountant is taken from the bottle and liberally dropped over the surface of the sections. A coverslip of suitable size (Fig. 12-40) is held at an inclined angle with a bent needle and slowly lowered so as to exclude all air bubbles. Then the edges of the slide are wiped roughly, and the slide is returned to the hot table shown in Fig. 12-39 to evaporate the solvent used for the resin.

Although this is the conventional method of operation, it is by no means the best. In particular, there is a tendency to have a higher concentration of solvent along the edges of the coverslip than in the center, and it also takes a surprisingly long space of time for all the solvent to

TABLE 3 | **Defects Appearing after Sections Have Been Cut**

Defect	Cause	Remedy	Method of prevention
Sections appear wrinkled.	1. Blunt knife used for cutting.	1. None.	1. Sharpen knife and cut new sections.
	2. Water used for flattening too hot, so that folds in sections fused into position.	2. None.	2. Watch temperature of water used for flattening.
	3. Sections unable to expand sufficiently: a. Because water used for flattening too cold. b. Because area of water too small.	3. None.	3a. Watch temperature of water used for flattening. b. Make sure that slide is clean, so that water flows uniformly over it.
Sections have bubbles under them.	1. Sections insufficiently flattened, so that air is trapped.	1. If sections still wet, reflood slide with water and reheat to complete flattening.	1. Check flatness of sections before draining slide.
	2. Air dissolved in water used for flattening comes out and is trapped under sections in drying.	2. If sections still wet, reflood slide with water, work out bubbles, and reheat to complete flattening.	2. Use air-free (boiled) water for flattening. Drain slide thoroughly and blot off excess moisture, Squeeze sections to slide.

TABLE 3 (Continued) | Defects Appearing after Sections Have Been Cut

Defect	Cause	Remedy	Method of prevention
Sections fall off slide.	1. Wax melted in flattening.	1. None.	1. Watch temperature of water used for flattening.
	2. Slide greasy.	2. None.	2. Use clean slides.
	3. Alkaline reagents dissolve albumen adhesive. (Sections start to work loose in course of staining or dehydrating.)	3. Transfer slides carefully to absolute alcohol. When dehydrated, dip in 0.5% celloidin in ether-alcohol. Then dip in 50% alcohol to coagulate celloidin. (This is not worthwhile unless sections are absolutely irreplaceable.)	3. See Gray's "Microtomist's Formulary and Guide" for other section adhesives not alkali-sensitive.
	4. Sections not flattened into perfect contact with slide.	4. None.	4. Sometimes caused by swelling of sections, which causes center to lift. Squeeze sections to slide and dry as rapidly as possible.
Sections contain fine opaque needles.	1. Imperfect removal of mercuric fixatives.	1. Return sections through proper sequence of reagents to water. Treat 30 min with Lugol's iodine, rinse, and bleach in 5% sodium thiosulfate. Restain.	1. Treat tissues in Lugol's iodine before embedding.
Sections contain black granules.	1. Long storage in formaldehyde.	1. None	1. Never store tissues in formaldehyde—always in paraffin blocks.

be removed. It is much better, if one can spare the time, to place a relatively thin coat of mounting medium on top of the slide and then to allow the solvent to evaporate from this on the surface of a hot plate. There is no risk that the slide will dry out, because the mountant will act as a varnish. Needless to say, in fine work it is necessary to cover the slide and hot plate with some dustproof cover while this is going on. The next day the slide is examined and, if it appears to be sufficiently varnished, the coverslip is placed on the surface. Then the whole slide is warmed, while maintaining steady pressure, above the softening point of the resin. The slide will be hardened as soon as it is cooled and may be cleaned and put away. This custom of evaporating the solvents from the surface of the slide rather than from the edge of the coverslip is considered old-fashioned nowadays, but there is no doubt that it produces a better and more durable slide than the more usual procedure.

It must not be imagined that, just because all these directions have been followed scrupulously, a perfect slide will result. There are nearly as many things that can go wrong with a section after it has been cut (see Table 2) as there are things that can happen in the course of cutting. Some of the more important of these are listed in Table 3, but it must be realized that no amount of written instruction can take the place of experience.

FROZEN SECTIONS

There are two circumstances under which paraffin sections cannot be used: first, where it is desired to preserve in the tissues some fatty material that would be dissolved out by the reagents used prior to impregnating; and, second, when speed is of primary importance, as in the production of quick sections from tumors for diagnostic purposes. In both these cases, recourse may be had to the method of frozen sections, in which material is rapidly frozen until it is of a consistency to cut. Frozen sections should not, however, be employed on any occasion when the normal processes of embedding can be used.

Choice of a Microtome. The type of microtome shown in Fig. 12-41 is so universally employed that it will be taken as the basis for the present discussion. It is essential in cutting frozen sections that the knife slice, rather than push, through the tissue. This slicing effect is produced by mounting the knife to swing through the object when the handle on top is turned. These microtomes are not so accurate, as to either the thickness of section cut or the repetition of this thickness, as the big Schantz shown in Fig. 12-18, but it is to be presumed that no one would cut

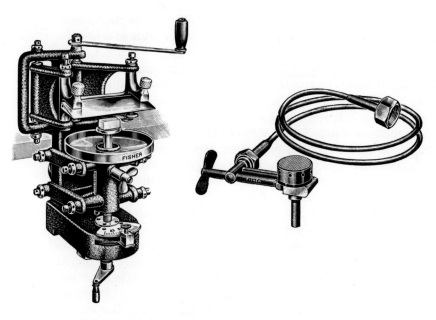

Fig. 12-41 | Spencer clinical microtome fitted for freezing.

frozen sections, in any case, if thickness and reproducibility were primary objectives. The method of freezing the object will be discussed after we have dealt with the question of supporting the material in a suitable medium. If there are available even a few moments in excess of the absolute minimum time required to cut without embedding, better results will be obtained if the object is smothered in several layers of:

Anderson's Medium:

Simple syrup	50 ml
80% alcohol	50 ml
White dextrin	15 g

Note: The dextrin is boiled to solution in the syrup and cooled; the alcohol is added slowly and with constant stirring. Simple syrup, which can be obtained from any pharmacist, is prepared by dissolving 5 lb of sugar in 2 pt of water.

Choice of Refrigerant. Blocks are nowadays almost invariably frozen with the aid of carbon dioxide, which is available very cheaply in large cylinders. The cylinder is connected through a needle valve to the object holder of the microtome, so that one has only to twist the valve to project a jet of supercooled carbon dioxide against the underside of the object holder.

Process of Cutting. The prime necessity for producing a good section is, of course, the availability of a sharp microtome knife. The nature, care, and sharpening of microtome knives have already been discussed. Assuming that more than the minimum time is available and that the method of Anderson is to be used, the following materials are required: a bottle of Anderson's syrup, which is conveniently kept in a balsam bottle; a pipette of the eyedropper type; and a dish of 70 per cent alcohol in which to receive the sections as they are cut. It is to be presumed also that the carbon dioxide cylinder has been attached to the tube leading to the microtome's freezing table and that a brief trial has been run to make sure that the gas is flowing satisfactorily.

About ½ ml of syrup is picked up with the pipette and placed on the freezing table of the microtome. A small jet of carbon dioxide is then turned on. Within a moment or two the gum will be seen to have congealed, and the carbon dioxide is turned off. The object to be sectioned is then (Fig. 12-42) placed on top of this congealed layer of gum, and more gum is poured over the surface. Care must be taken that a layer of

Fig. 12-42 | Applying Anderson's medium to tissue about to be frozen.

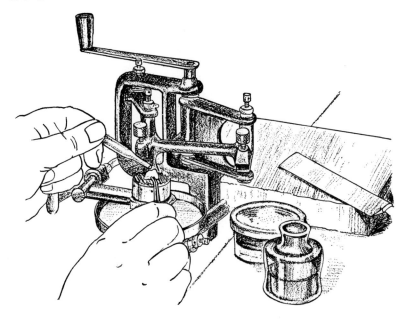

uncongealed gum lies between the object and the congealed gum, or it may loosen. The carbon dioxide is again turned on, and, as soon as the gum covering the object is seen to be congealing, a little more syrup is poured on the surface so that the object is thoroughly covered. It will be seen from the figure referred to that, as a matter of convenience, the knife has not yet been placed in its holder while these preliminary operations are going on. As soon as the knife has been inserted, an experimental cut is taken across the top of the material with the knife, and the block is then shaved down until the specimen is reached. The device that controls the thickness of the section to be cut is then set to whatever thickness has been decided upon. It is recommended that beginners not attempt to cut sections much less than 20 μ in thickness by this method, and often sections of 30 μ are sufficiently good for diagnostic purposes. It is then necessary only to continue to pull the handle until the object starts to cut, while observing the nature of the sections. If the sections crumble under the action of the knife, while the gum melts instantly on contact with it, it may be presumed that the block has not been frozen sufficiently hard; the carbon dioxide may again be turned on for a few moments and another cut taken. It will take only a moment or two to establish the optimum condition under which only slightly curled sec-

Fig. 12-43 | **Removing section from knife.**

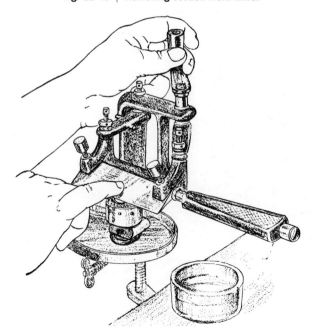

tions appear on the blade of the knife. As the blade, however, is likely to have become soiled, it is now washed with a drop of warm water to remove the dried gum and then used to cut as many sections as are required. As each section is cut, it must be removed from the blade of the knife to the dish of 70 per cent alcohol shown in the figure. The majority of people working under pressure use their little finger (Fig. 12-43) for the removal of the section, although a number of very competent technicians prefer to use a brush for this purpose. As soon as the section has been removed, it is transferred to the 70 per cent alcohol, where the gum will soon be dissolved.

Cutting without support is much more difficult and should be confined to homogeneous blocks of tissue. In this case the tissue, either fresh or fixed in formaldehyde, is trimmed so that a straight edge will be opposed to the edge of the knife. A drop of water is placed on the surface of the freezing table, the gas is turned on, and the specimen is pressed firmly to the water while it is frozen in place. The finger is then removed and the block watched until the line of frost on the side just reaches the top. Cutting should start at once, using a rather slow even stroke. The correct degree of freezing is vital to success but can be learned only by experience.

Mounting and Staining Frozen Sections. Fairly thick frozen sections can be handled with section lifters and passed through any of the stains described in Chapter 3. Sections of fatty tissue are always handled in this manner and should be stained by the method discussed in Chapter 7.

If complex staining techniques are to be applied, however, it is better to fix the section or sections to a slide. One of the best techniques involves two solutions:

Gravis's Adhesive:

Water	100 ml
Agar	0.1 g
Camphor	0.1 g

Dissolve the agar in boiling water. Filter and add the camphor to the hot filtrate.

Zimmerman's Lacquer:

Absolute alcohol	50 ml
Anhydrous ether	50 ml
Pyroxylin U.S.P.	
or Parlodion	0.5 g
Gum mastic	0.1 g

The first solution is used exactly as water is used to flatten a paraffin section. The amount of warming necessary depends on the degree of wrinkling of the section but is usually less than for paraffin ribbons. As soon as the slide has been drained, absolute alcohol is dropped gently on the section from a pipette, left for a few moments, and then drained off. A few moments now suffice to dry the sections, which become opaque. A piece of lens paper is then laid over the slide, and each section is pressed firmly with the finger. The slide is now quickly dipped in absolute alcohol, drained, and dipped in Zimmerman's solution. This last is allowed to evaporate from the surface, leaving behind a coat that will hold the section in place through subsequent staining techniques. Any of the stains discussed in Chapter 7 may be used save those involving differentiation in absolute alcohol, which would dissolve the lacquer. The slides are treated exactly as though they carried paraffin sections except that the final transfer should be from 95 per cent alcohol to terpineol to xylene. Even terpineol has a slight solvent action on the slide, which should remain in it, therefore, only long enough to remove the alcohol.

SUGGESTED ADDITIONAL READING

Johansen, D. A.: "Plant Microtechnique," New York, McGraw-Hill Book Company, 1940.

Richards, O. W.: "The Effective Use and Proper Care of the Microtome," Buffalo, N. Y., American Optical Company, 1949.

Steedman, H. F.: "Section Cutting in Microscopy," Oxford, Basil Blackwell & Mott, Ltd., 1960.

CHAPTER 13 | Cleaning, Labeling, and Storing Slides

There is not much point in going to all the trouble and difficulty of making a good microscope slide unless one is also prepared to finish, label, and store it properly. There is a great deal of difference between a slide that is merely left after the coverslip has been placed on it with a label stuck roughly on the end and one that has been properly finished and properly labeled. This, indeed, is the principal difference between the professional slide purchased from the biological supply house and one that is turned out by the average beginning student.

There are two stages to finishing a slide. The first of these is to clean from the outside all unwanted mountant and to polish the glass. The second is to attach to it a label that is both neat and permanent. It is easy to remove the unwanted gum mountant from wholemounts if one is using either the medium of Farrants or the medium of Berlese, for it is necessary only to wipe very gently with a damp cloth until the surplus has been removed. If far too much of the medium has been used, so that there is a large exudate around the edge of the coverslip, it is usually better to work in two stages, that is, to remove about half of this exudate one day and the remainder the next day. The reason for this is that these gums harden only on the outside, and, if the whole of the surplus is washed off at one time, there is a grave risk of displacing the coverslip. A somewhat different procedure is employed when one is cleaning a slide that has been mounted in the medium of Gray and Wess. This material dries to a tough pellicle which is not water-soluble but is easily removed with a knife. The sharp point of a scalpel is run either around or along the edge of the coverslip, and then the surplus hardened pellicle is picked off as a single sheet. The cut must extend all the way through to the glass, or the coverslip, with the object attached to it, may come off. However, this is not a permanent catastrophe because one can easily remount the coverslip with its adherent object in the same medium.

In either case, after the surplus medium has been removed, the slide is left for at least 2 days to harden, dipped very briefly in a finger bowl of a cold soap solution, and then dried and polished. The slide must not be left in the soap long enough to loosen the coverslip, or the coverslip is likely to be removed when polishing the slide.

Cleaning wholemounts made in resinous media is a rather different procedure. It is necessary to wait until the resin has completely hardened, hastened where necessary by exposure on a hot plate. Then as much of the surplus as can be removed is scraped off with a dull knife. A small lump of absorbent cotton saturated in 96 per cent alcohol is used to rub off the remainder of the resin. It is not safe to try to clean a balsam slide with benzene or xylene, for the coverslip will inevitably be loosened by this method. As soon as the surplus resin has been dissolved away, the slide is dipped immediately into a warm soap solution and then polished. This leaves a very brilliant finish and an entirely clean slide.

An additional reason why slides should always be washed in soap and water after they are mounted is that there is no known method by which a gummed paper label can be made to adhere to greasy glass. More slides are rendered useless through the loss of their labels than from any other defect. Almost everyone today buys pregummed paper labels, so that it is scarcely worthwhile in this text to give formulas for label adhesives. There is, however, a very definite technique by which the label may be made to stick most readily. This is to lick the *upper* surface of the label, moisten thoroughly the lower surface, and then, when both sides are fully expanded, press it firmly on the glass. Labels attached in this manner have remained for more than twenty years in the author's collection without becoming detached. One may then wait until the label is dry before printing neatly on it with waterproof India ink the name of the specimen or, if one is dealing with a large series of slides, one may write the label first in waterproof India ink and, having left it at least 24 hr to dry, attach it by the method indicated.

Even if the label is attached by the method indicated, it is always best, in the case of valuable slides, to write a brief label with a writing diamond on the glass underneath the label.

There is only one absolutely permanent method of labeling a microscope slide, and that is to use a slide of which about an inch at the end has been ground to a rough surface. After the slide is finished and cleaned, the label is written on this ground surface either with a soft pencil or with waterproof India ink. Then a drop of balsam is placed on the label before attaching a coverslip over the surface of it. The objections to this method are that it is very expensive—the slides cost four or five times as much as ordinary slides—and also that there is a grave risk that the coverslip will be broken by someone placing the clip of the microscope stage on top of it.

The proper storage of microscope slides is just as necessary as proper labeling. Slides should always be stored in the dark and in as cool a place as possible because none of the stains used in biology is absolutely permanent. Slides may be stored either vertically on edge in grooves in a tray of a cabinet of the type shown in Fig. 13-1 or lying flat on the bottom of a tray in the type of cabinet shown in Fig. 13-2. There are arguments for and against both methods of storage. Slides that are stored vertically occupy much less space than those that are stored flat. The former method is not very suitable for wholemounts for the reason that the interior of the slide never dries and, after a slide has been stored for some years in a vertical position, the object will be found to have dropped down to the edge of the coverslip and become embedded in the hardened balsam, from which it is very difficult to detach it later. There is, however, no objection to the storage of sections on edge, and many thousands of slides stored by this method can be kept in the space occupied by only a few hundred when they are stored flat.

Another method of storing slides, which is particularly useful in the case of serial sections involving many dozens of slides in each series, is to take two 3- by 5-in. index cards and to cut about an inch from the long edge of one of them. They are stapled together, making a series of

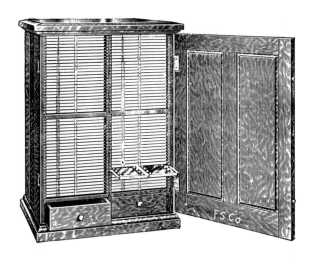

Fig. 13-1 | Slide-storage cabinet with vertical grooves.

194   The Preparation of Microscope Slides

Fig. 13-2 | Slide-storage cabinet for flat storage.

pockets into each of which a slide may be inserted. The full data connected with the slide may be written on the card, which may be stored in any of the ordinary card-filing cabinets that are available.

There are many types of small boxes made that have grooves to hold the slides. It is to be remembered with these that the slide must be kept in a horizontal position if it is a wholemount.

There is a considerable case to be made for ringing the edges of coverslips with a colored varnish, not only to improve the appearance of the slide but also to increase its permanence. This procedure is described in full in Gray's "Microtomist's Formulary and Guide," and space does not permit it to be given in the present text.

PART THREE | Specific Examples of Slide Making

EXAMPLE 1 | Preparation of a Wholemount of a Mite by the Method of Berlese

The use of the name Berlese in the heading of this example is less an injunction to employ the mounting medium of that writer than it is a tribute to the method of collecting small arthropods, which he introduced. This method uses the Berlese funnel. This device is a double-walled funnel, between the walls of which warm water may be placed and maintained at any desired temperature by applying a small flame to a projecting side arm. The temperature is not critical, so that no thermostatic mechanism is provided, but a thermometer may be inserted and used to read the temperature at intervals. A circle of wire gauze with a mesh of about 1/16 in. is placed at the bottom of the inner glass funnel, and the material that is to be searched for mites is placed loosely on this gauze. The lower end of the glass funnel is attached with modeling clay to a tube containing whatever medium is being used for the collection of the specimens. If the specimens are to be stored rather than mounted at once, 96 per cent alcohol may be placed in the tube, and it is unnecessary to seal it at the base of the funnel. If, however, the specimens are to be mounted at once in Berlese's medium, in which much better mounts can be prepared from living than from preserved material, the tube must contain water and be sealed to the funnel in order to prevent the more active forms from working their way out of the tube.

After the moss has been placed in position, a small lamp of not more than 15 watts is mounted in any kind of a reflector some distance above the material. The animals in the material, therefore, find themselves surrounded by heat at the sides and plagued with light from above. As all these animals are violently photophobic, they tend to move automatically toward the lowest point of the moss, from which they drop down the funnel into the tube. By this means it is possible in 10 or 15 min to collect the whole fauna from a large handful of any organic material that would, by any other means, take several hours to search. The use of the funnel is not confined to moss but may be applied to hay, straw, shredded bark, or indeed any other material from which small arthropods are customarily collected. The only difficulty in using this equipment is in preventing the heat from becoming too great. Some people use so large a lamp above and so high a temperature around the edges that many small arthropods are killed before they have time to fall into the trap laid for them. The author has found that the water between the walls of the funnel for most uses should be at a temperature of 30 to 40°C, and the lamp above should under no circumstances raise the surface temperature of the material above 60°C. These temperatures are for a moderately dry moss sample and may be exceeded greatly when one is dealing with dry material, such as straw. Wet moss of the sphagnum type, however, requires lower temperatures if it is to be examined successfully.

If permanent mounts are to be made for record purposes of all the small invertebrates that may be found in the moss sample, it is necessary to make adequate preparations to receive the animals while the moss is being treated. Two kinds of gum mountants are desirable: (1) a high-refractive-index medium like Berlese's for the very heavy-walled forms, such as the Oribatid mites and the pseudoscorpions, and (2) a low-refractive-index mountant like Gray and Wess's for the thinner-walled forms, such as the Tyroglyphid and Gamasid mites. This last medium is also suitable for Thysanura and for Collembola. Thick-walled beetles and fleas, if they are to be made into microscope slides, had better be treated with alkali and should be accumulated for this purpose in a tube of 96 per cent alcohol. This process is described in detail in Gray's "Microtomist's Formulary and Guide." It consists usually of soaking the specimens after rehydration in 10 per cent potassium hydroxide. Excessive swelling may be controlled by enlarging the anus with a needle and by putting holes in unimportant parts of the head and thorax.

If one is dealing with a sphagnum moss, it is also possible that a number of crustaceans, particularly Cladocera and Ostracoda, are likely to be found. These are better mounted in glycerin jelly in the manner also described in Gray's "Microtomist's Formulary and Guide" and should be transferred, as soon as they are found, to 30 per cent alcohol, where they will die with their appendages extended. They should not be permitted to remain in this weak alcohol for longer than is necessary to kill them. Then they should be transferred to 96 per cent alcohol. A large number of nematode worms are likely to turn up; these cannot be mounted by any of the methods described in this book, and again reference should be made to the author's larger work. A tube of some fixative, a supply of clean 3- by 1-in. glass slides, and a number of coverslips should be provided to receive any small annelids which may be found in the moss, and which must be fixed, stained, and mounted at once.

When all is ready and observation shows that no more forms are falling through the Berlese funnel, the collecting tube beneath it is inspected to see roughly what has been gathered. If there are a great number of Gamasid mites, or active insects, it is necessary to open gently a portion of the tube by pushing away the modeling clay with the thumb and to let a minute drop of ether run down inside. When this has been done, the tube is removed, and the contents are tipped out into a petri dish or similar container, ready to be mounted. The specimens are best collected with the aid of a fine brush that has been moistened in water.

A mite or similar form that is to be mounted is selected, picked up on the tip of the brush, and transferred to a drop of whichever of the two gum media has been selected. As little water as possible should be transferred with it, and the mite should be pushed under the surface of the gum with the point of a needle. The mount is inspected under the low power of the microscope, and, if any large quantity of air has been carried in with the mite, the bubbles are released with the aid of a fine needle and allowed to come to the surface before the coverslip is laid gently into place. The drop of gum should be large, and no endeavor should be made to press the coverslip down. If a reasonably thick layer of mountant is left, almost any small arthropod will spread its legs as in a textbook diagram before dying and will remain in this form indefinitely. The finishing of these slides has already been discussed and may be carried out at any time after they have been made.

SUMMARY

1. Erect Berlese funnel.
2. Transfer small arthropods to drop of mounting medium on slide.
3. Add coverslip.

EXAMPLE 2 | Preparation of a Wholemount of Pectinatella Stained in Grenacher's Alcoholic Borax Carmine

Although this description applies to the animal named, it may be used equally well for any other fresh-water bryozoan or, as a matter of fact, for any small invertebrate of about the same size and consistency. Pectinatella has been chosen only for the reason that it has the habit of turning up on the walls of the aquaria in the author's laboratory. Profitable hunting grounds, if they have to be sought, are the undersides of the leaves of large water plants and the surfaces of branches of trees that have fallen into the water but have not yet had time to decay. An old trick of European collectors was to lower a length of rope into a pond in which Bryozoa were known to occur and to leave it there for the summer. It was astonishing how frequently, when these ropes were pulled up again in the fall, they were found to be covered with colonies of Bryozoa.

However the Bryozoa are obtained, it is necessary first that they be narcotized. The material on which they are living is cut up and placed on the bottom of a finger bowl of aquarium or pond water. Distilled water and tap water are lethal to these forms. There should not be so many specimens that they touch each other on the bottom of the finger bowl, and the finger bowl itself should be completely filled with water. The fresh-water Bryozoa are a little sensitive to heat and may not respond well to the high temperatures found in some laboratories. In this case it is well to put the finger bowl containing the specimens in an icebox or electric refrigerator, preferably one held at about 10°C, and to leave it there overnight. Then the specimens may be brought out and narcotized before they have time to suffer from the increasing temperature.

The author prefers to use menthol, which is both cheap and easy to obtain. The menthol is sprinkled on the top of the water in the specimen jar. For an ordinary finger bowl, about 1 g of menthol will be sufficient. There is no means of foretelling how long it will take the specimens to become narcotized; therefore, they should be observed at intervals until they no longer are seen to be contracting. However, this may not be due to narcotization, so that some very delicate instrument—a hair mounted in a wooden handle is excellent—should be used to test narcotization by pushing the individual polyps. If, on receiving a push, they contract sharply, it is evident that no narcotization at all has taken place, and the amount of menthol that has been sprinkled on the surface should be increased greatly. If, on being pushed with a hair, they contract slowly, it is evident that they are partly narcotized. One must be careful not to disturb them further for at least 10 min, for if they contract in a narcotized condition they will not expand again. The right stage for killing has arrived when no amount of shoving with a hair will persuade the specimens to contract and an examination under a binocular microscope shows that the ciliary action on the lophophore has not stopped. A rubber tube is used to siphon carefully from the finger bowl enough water so that the remaining layer just covers the specimens. Then the finger bowl is filled with 4 per cent formaldehyde, covered, and put to one side.

A careful distinction must be made between a killing agent, such as formaldehyde, and hardening and fixing agents. In the present instance it is quite unnecessary, since a stain containing in itself an adequate mordant is to be used, to employ any fixative that will combine with the proteins of the specimens, but it is necessary to harden them in order that they may withstand the treatment to which they will be subjected in staining and dehydration. Four per cent formaldehyde hardens very slowly, and it is suggested that next the specimens be passed to alcohol for the hardening process.

It is desirable, however, that they be flattened, before hardening, into the shape that they will be required to assume after mounting. It is to be presumed that the purpose of making a microscope slide is to study the object that has been mounted; the depth of focus of microscope lenses is so slight that only relatively thin objects can be studied. It is extraordinary how frequently this simple principle is overlooked or how frequently people endeavor to flatten the object after it has been mounted in balsam and is almost invariably so brittle that it will break up during the flattening process. Five minutes' work in arranging the parts before hardening makes all the difference between a first-class and a second-class mount. To arrange and flatten the objects for hardening, the 4 per cent formaldehyde is replaced with water—a matter of convenience—and then the first specimen to be treated is selected. This specimen is removed to a finger bowl of clean distilled water, where it is examined thoroughly to make sure it has no adherent dirt. The object is flattened by hardening it between two slides, but it will be obvious that if it is just pressed between two slides it will be squashed rather than flattened. Anything may be used to hold the two slides apart, although in the present instance a very thick No. 3 or two No. 2 coverslip would give about the right separation. Therefore, at about an inch on each side of the center of a glass slide is placed a thick No. 3 coverslip, which may be held in place by the capillary attraction of a drop of water. The specimen is picked up from the water with a large eyedropper type of pipette and placed in a large volume of water on the slide. It is then easy to arrange the parts with needles, but it is difficult to lower a second slide without disarranging these parts. An alternative method is to place the slide with its coverslip in the finger bowl with the specimen, to arrange its parts under water, and to place the second slide on top. Whichever process is adopted, the slides are tied or clipped together and transferred to a jar of 96 per cent alcohol, where they may remain for a week or until next required. Each specimen is treated in this manner. It is better not to try to flatten two or three specimens on one slide.

When it is time to continue mounting the specimens, the slides are placed in a finger bowl of 96 per cent alcohol before cutting the cords or removing the clips that bind them together. Getting the two slides apart without damaging the specimen is not easy, particularly if the specimen tends to stick to one or the other of the slides. The simplest method is to insert the blade of a scalpel into the gap between the slides and twist it slightly to see whether or not the specimen is free. If the specimen shows signs of sticking to one slide, the other may be removed and the specimen washed from the slide to which it is stuck with a jet of 96 per cent alcohol from a pipette. If it shows signs of sticking to both slides, it is still possible to free it from both by projecting a jet of 96 per cent alcohol between them. Each slide is treated in due order until one has accumulated all the flattened specimens in a dish of 96 per cent alcohol. It must be understood that these specimens have been hardened flat, so that no amount of subsequent treatment will ever swell them out again or prevent them from remaining in the required position.

It is recommended, if there are several specimens to be handled, that a series of the little cloth-bottomed tubes shown in Fig. 10-1 be used. The only alternative is to handle each specimen with the aid of a section lifter, with the consequent risk of damage. Although not nearly so satisfactory, it is also possible, at least for the process of staining and dehydration, to place all the specimens in a small vial in which the different fluids used may be placed successively.

A wholemount of this type is best stained in carmine, and the choice would lie between Mayer's carmalum and Grenacher's alcoholic borax carmine; the author's preference is for the latter. The preparation of the latter stain, the formula for which is given in Chapter 7, does not present any difficulty, but it should be noted that a differentiating solution of 0.1 per cent hydrochloric acid in 70 per cent alcohol will be required. Adequate supplies of this should be available before one starts staining.

The specimens are now passed from 96 per cent alcohol to 70 per cent alcohol. Naturally they will float, but as soon as they have sunk to the bottom it may be presumed that they are sufficiently rehydrated. Either the cloth-bottomed tube containing them may be transferred to the dish of stain or the 70 per cent alcohol may be poured out of this tube and stain substituted for it. Two of the advantages of this stain are that it is relatively rapid in action—very few specimens will not be stained adequately in 5 to 10 min—and it does not matter how long the materials remain in it. It is, therefore, often convenient to leave the specimens in stain overnight and to start differentiation the next morning. Either they are removed to differentiating solution or, alternatively, the stain is poured off and the differentiating solution substituted for it. In the latter case, three or four changes will be required, owing to the necessity of leaving some stain in the bottom of the tube to avoid pouring the specimens out with it. Indeed, unless the operator is experienced, it is safer to shake the tube, so as to distribute the specimens thoroughly in the stain, and then to tip everything into a large finger bowl of differentiating solution from which the specimens may be picked out later and transferred to a new tube of differentiator. It is tragically easy, in pouring off the stain, to pour specimens with it down the sink. As soon as the stain has been washed off with the differentiating solution, a single specimen should be transferred to a watch glass and examined under a low power of the microscope. It is more than probable that little differentiation will be required, so a simple rinse may be adequate. It is difficult to judge the exact degree of differentiation required, and it must be remembered that the object will appear darker after clearing than it does in the differentiating solution. The internal organs should be sharply demarcated when the outer surfaces of the specimen are relatively free from stain. This may be judged in Pectinatella by placing a coverslip on the specimen and examining one of the branches of the lophophore under the high power of the microscope. Differentiation may be considered complete when only the nuclei in the cells of the lophophore are stained. The specimens are washed in four or five changes of 70 per cent alcohol to remove the acid before they are placed for at least a day in 96 per cent alcohol as the first stage of dehydration. Next they should be transferred to two changes of at least 6 hr each in a considerable volume of fresh 96 per cent alcohol and cleared. Absolute alcohol is not necessary if one is using terpineol as a clearing agent.

There is some danger, if specimens are transferred directly from 96 per cent alcohol to a fluid as viscous as terpineol, that they will become distorted by the violent diffusion currents. This may be readily avoided in the following manner: A fairly wide (about 1 in.) glass vial is filled about half full with terpineol. Ninety-six per cent alcohol is poured very carefully down the side of the vial in order to float a layer of alcohol on top of the terpineol. The specimens are now dropped into the alcohol. Naturally they sink through it, coming to rest on the surface of the layer of terpineol into which they sink slowly without any strong diffusion currents. They will be seen to have sunk to the bottom after a little while, but there will still be a quantity of alcohol diffusing upward from them. As soon as these diffusion currents have ceased, the alcohol should be drawn from the top of the tube with a pipette and the specimens transferred to clean terpineol. When they are in fresh terpineol, they should be examined carefully under a microscope to make sure that they are glass-clear without the least trace of milkiness. If they appear slightly milky, either they have been dehydrated insufficiently or the alcohol used for the preparation has become contaminated with water. In either case, before being transferred to a tube of fresh alcohol for complete dehydration they must be transferred back into terpineol in the manner described. It is a sheer waste of time to endeavor to prepare a balsam mount from a specimen that is not perfectly transparent in the clearing medium.

When all the specimens are in natural balsam, have ready some clean slides, some clean ¾-in. circular coverslips, and a balsam bottle containing natural balsam. The author's preference for the natural balsam rather than a solution of this material in some solvent has been explained previously. For example, a drop of the natural balsam is placed on each of six slides, and then, one at a time, six specimens are lifted from the terpineol and placed on top of the balsam. The specimens will sink through the balsam very slowly, so that these six slides should be pushed aside while the next six slides have drops of balsam put on them, and so on. As soon as a specimen has sunk to the bottom of the drop of balsam, a coverslip is held horizontally above, touched to the top of the drop, and then pushed down with a needle until the specimen is flattened firmly against the slide. Since these specimens have been properly hardened and flattened, there is no risk of their being damaged by drying the mount under pressure, so a clip can be applied (see Fig. 10-4) and the slides placed on a warm table to harden. Each is cleaned, finished, and labeled in the manner suggested previously.

SUMMARY

1. Narcotize with menthol.
2. Kill with 4 per cent formaldehyde.
3. Harden selected polyps compressed between slides in 95 per cent alcohol.
4. Stain 1 to 24 hr in Grenacher's alcoholic borax carmine.
5. Differentiate in acid 70 per cent alcohol until pink.
6. Dehydrate, clear in terpineol, and mount in balsam.

EXAMPLE 3 | Preparation of a Wholemount of 33-hour Chick Embryo, Using the Alum Hematoxylin Stain of Delafield

Fertile eggs are relatively easy to obtain and should be incubated at a temperature of 103°F for the required period of time. The term *33-hr chick* is relatively meaningless since the exact stage and development that will have been reached after any given time in the incubator depend on the temperature of the latter, on the temperature at which the egg was stored prior to its incubation, and even on the age of the hen. It is desirable, therefore, if any very specific age of development is required, to start a series of eggs in the incubator at 3- or 4-hr intervals and then fix and mount these at the same time.

For the removal of the embryos from the egg, there is necessary first a series of finger bowls or any circular glass dishes 5 to 6 in. in diameter and 2 to 3 in. in depth, a number of syracuse watch glasses, a large quantity of a 0.9 per cent solution of sodium chloride, a pair of large dissecting scissors, fairly fine forceps, a pipette of the eyedropper type, some coarse filter paper, and a pencil.

No very great accuracy is required in making up the normal salt solution, and any percentage between 0.7 and 1 will be sufficient for the purpose in mind. Although it is customarily specified that the temperature of the solution be 102 to 103°F, anywhere within 10° on either side of these figures is relatively safe. The egg is removed from the incubator and placed in one of the finger bowls, to which is added the warm normal saline solution until the egg is totally immersed. If the operator is rather skilled, it is possible, of course, just to break the egg into the warm saline solution as though one were breaking it into a frying pan, but it is recommended that the inexperienced prepare several hundred wholemounts before they endeavor to do this. The method by which the inexperienced can be assured of getting a perfect embryo on every occasion is first to crack open the air space which lies at the large end of the egg and to let the air bubble out through the warm saline solution. This permits the yolk to fall down out of contact with the upper surface of the shell, which may be removed with blunt-nosed forceps, working from the air space toward the center. Again the matter of practice is involved, for a skilled operator can remove this shell in large portions, whereas the inexperienced must work very carefully to avoid puncturing the yolk. If the yolk is punctured, it is much simpler to throw the egg away and start with another one. After about half the shell has been removed, it will be found relatively easy to tip the yolk, with the embryo lying on top of it, out into the saline solution and remove the shell.

The next operation is to cut the embryo from the yolk by a series of cuts made well outside the terminal blood vessel, which marks the limits of the developing embryonic structures. To do this with success requires more courage than experience. Just as soon as the vitelline membrane

Figs. E3-1, E3-2, and E3-3 | Chick embryo prepared by the method described in this example. Fig. E3-1. An excellent slide. Fig. E3-2. A good slide ruined by the application of a dirty coverslip. Note dust granules at 1 and 2. Fig. E3-3. A complete failure. The wrinkles at 3 show that the embryo was not properly attached to filter paper before fixing. The drawn-out diffuse myotomes at 4 and the absence of head development indicate that the embryo had died some time before fixation. The crack in the dark background at 5 indicates that the yolk was imperfectly removed before fixation.

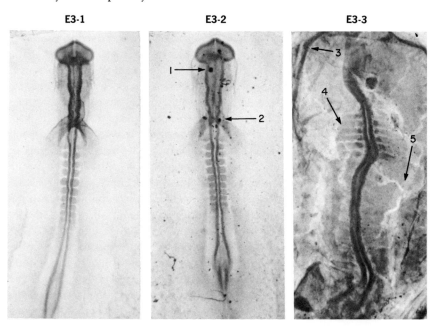

is punctured, the yolk starts squirting out through the hole, rendering the fluid milky, so that one can no longer see the embryo. The smaller the hole cut, the more violently does the yolk squirt out, so the larger the scissors that can be employed, the more easily will the embryo be removed. The easiest method is to take blunt forceps in the left hand and with it grip the extraembryonic areas of the chick well outside the sinus terminalis. A certain amount of drag is placed on it so that the vitelline membrane is wrinkled, and then with a large pair of scissors a transverse cut is made directly away from the operator about ⅓ in. outside the sinus terminalis on the side of the embryo opposite that which is held by the forceps. This initial cut should be at least ¾ in. long and should be made firmly. Two cuts, at right angles to the first, should be run on each side of the embryo. The part gripped with the forceps should be released, and the free edge, where the first cut was made, gripped so that the embryo can be folded back away from the yolk. It is now relatively easy to sever all connection between the embryo and the underlying materials by a fourth cut. The embryo, held by the forceps in the left hand, will now be floating free in the saline solution. The embryo is much stronger than it looks and will not be damaged, provided the tips of the forceps are kept under the saline solution.

The embryo must be transferred to a clean saline solution, preferably in another finger bowl. This transfer may be made either with a very wide-mouthed pipette of the eyedropper type or by scooping it up in a smaller watch glass with plenty of saline solution and transferring it to the fresh solution. Here it should be picked up again by one corner with the forceps and waved gently backward and forward to remove the adherent vitelline membrane as well as such yolk as remains. At this stage the embryo should be examined to make sure that the heart is beating and that it is in fit condition for fixation.

The embryo is now scooped out on one of the syracuse watch glasses with as little water as possible. Next it is necessary to persuade it to flatten on the bottom in an upside-down position, that is, so that the portion of the embryo that was previously in contact with the yolk is now directed toward the operator. To determine which side of the embryo is uppermost requires considerable practice unless the primary curvature of the head toward the right has already started. The best point of examination is the heart, which lies, of course, on the lower surface of the embryo. Having maneuvered the embryo in the saline solution in the watch glass until it is in the upside-down position required, the water should be drained off with the aid of a pipette, which is run rapidly with a circular motion around the outside of the blastoderm while the water is drawn up. As experience will soon show, any attempt to drain the water up a stationary pipette will result in the embryo being drawn out in the direction in which the water is being sucked. A little practice in running the pipette around and around the outside of the blastoderm and about a millimeter away from it will enable the operator to strand the embryo perfectly stretched in all directions. Under no circumstances whatever should a needle be used in an endeavor to arrange the embryo because the point will adhere to the blastoderm, from which it cannot be detached without damage. If the embryo is not flattened and spread out satisfactorily, it is necessary only to add a little clean saline solution with a pipette and repeat the operation.

A piece of coarse filter paper or paper toweling is cut into a rectangle of such size that it will drop easily into a syracuse watch glass. An oval or circular hole is cut in the middle of this (done most easily by bending it in two and cutting a semicircle) of such a size as will cover exactly those areas of the embryo which are to be retained. That is, if the embryo alone is required, the hole may be relatively small, while if it is desired to retain all the area vasculosa with its sinus terminalis, the hole must be correspondingly enlarged. The hole must not be larger, however, than the blastoderm removed from the egg because the next operation is to cause the unwanted extraembryonic regions to adhere to the paper, leaving the embryo clear in the center. By this means alone will the embryo be prevented from contracting and distorting when fixative is applied to it. Such data as are pertinent may be written on the edge of the paper rectangle in pencil. The paper is then dipped in clean saline solution. If the saline used has already become contaminated with egg white, a sharp puff of air should be directed at the hole to make quite certain that a film of moisture does not extend across it since the bubbles so produced always disrupt the embryo if this film is left. The rectangle of filter paper is now dropped on top of the stretched embryo in such a manner that the embryo does not become distorted. This is, in point of fact, a great deal easier than it sounds, although a few false trials may be made by the beginner. The author's procedure is to place one end of the rectangle on the edge of the watch glass nearest him, taking care that it does not touch the blastoderm, and then to let the paper down sharply. The edges of the blastoderm must be in contact with at least two-thirds of the periphery of the hole if it is to remain stretched. As soon as the paper has been let down, the end of a pipette or a needle should be used to press lightly on the edges of the paper where it is in contact with the blastoderm to make sure that it will adhere.

The embryo is now ready for fixing. The choice of a fixative, naturally, must be left to the discretion of the operator. The author's preference, where hematoxylin is to be employed for staining, is for a mercuric mixture, such as the solution of Gilson. If much embryonic work is to be undertaken, reference should be made to Gray's "Microtomist's Formulary and Guide" for the formula of Gerhardt's fixative, which is better. The disadvantage of the customarily used picric acid formulas is that they interfere seriously with subsequent staining by hematoxylin. The fixative should be applied from an eyedropper type of pipette in the following manner: A few drops are first placed on the center of the embryo, so that a thin film of fixative is spread over it. After a moment or two a little more may be added with a circular motion on the paper surrounding the embryo. Again the paper should be pressed on the periphery of the blastoderm with a needle or the end of the pipette to make sure that the adherence is perfect, and the whole should be left for a moment or two before being shaken gently from side to side to make quite certain that the embryo is not sticking to the watch glass. If it is sticking, the end of the pipette containing the fixative should be slid under the edge of the paper and a very gentle jet or fixative used to free the embryo. As soon as the embryo is floating freely in fixative, the syracuse watch glass may be filled up with fixative and placed to one side, while the same cycle of events is repeated with the next embryo. After about 10 min in the fixative, the paper may be picked up by one corner and moved from reagent to reagent without the slightest risk of the embryo becoming either detached or damaged. Care, of course, must be taken not to pick up the paper with metal forceps unless the instrument has been waxed first because the mercuric chloride in the fixative will damage the metal. It is the author's custom to leave the embryos in the watch glasses for about 30 min before picking them out and transferring them to a large jar of the fixative, preferably kept in a dark cupboard. The total time of fixation is not important but should be not less than 1 day nor more than 1 week. When the embryos are removed from the fixative, they should be washed in running water overnight and can then be stored in 70 per cent alcohol for an indefinite period.

When a batch of embryos is ready to stain, it is necessary only to take them from 70 per cent alcohol back to distilled water until they are thoroughly rehydrated and then transfer them to a reasonably large volume of Delafield's hematoxylin, where they may remain overnight. The gravest mistake that can be made in this type of staining is to stain initially for too short a period. The result is that the outer surface of the embryo becomes adequately stained while the inner structures do not, but this defect is difficult to detect until the embryo is finally cleared for mounting. When the embryos are removed from the stain, at which time they should appear a deep purple, they should be transferred to a large finger bowl of distilled water, where they are rocked gently backward and forward until most of the stain has been removed from the paper to which they are attached. Each embryo should then be taken separately and placed in 0.1 per cent hydrochloric acid in 70 per cent alcohol. The color will immediately start to change from a deep purple to a pale bluish pink, and the embryos should remain in this solution until, on examination under a low power of the microscope, all the required internal structures appear clearly differentiated. Most people differentiate too little, forgetting that the pale pink of the embryo will be changed back to a deep blue by subsequent treatment and that the apparent color will also increase in density when the embryo is cleared. No specific directions for the extent of the differentiation can be given beyond the general advice to differentiate far more than you would anticipate to be necessary (Fig. E3-1 is a good guide). After the embryos have been sufficiently differentiated, each one should be placed in tap water, rendered alkaline with the addition of sodium bicarbonate. Here it should remain until all the acid has been neutralized and the embryo itself has been changed from a pink back to a deep-blue coloration. It may then be dehydrated in the ordinary manner through successive alcohols. It is the author's custom to remove it from its paper only when it is in the last alcohol and before placing it in the clearing agent. Some people place it in the clearing agent attached to its paper and remove it only before mounting. Any clearing reagent may be tried at the choice of the operator; the author's preference for chick embryos is terpineol, which has the advantage of not rendering these delicate structures so brittle as many other reagents. The cleared embryo is mounted in balsam.

SUMMARY

1. Incubate fertile eggs at 103°F for 33 hr.
2. Remove shell under surface of warm 0.9 per cent sodium chloride.
3. Detach embryo from surface of yolk with bold cuts and wash in clean saline solution.
4. Strand embryo in watch glass.
5. Cut hole slightly larger than embryo in filter-paper rectangle, soak paper, and drop it over embryo.
6. Fix in Gilson's fluid 1 to 6 days.
7. Wash in running water overnight.
8. Stain overnight in Delafield's hematoxylin.
9. Differentiate in acid 70 per cent alcohol until internal anatomy is shown clearly.
10. Treat with alkaline tap water until blue.
11. Dehydrate.
12. Detach embryo from paper and clear in terpineol.
13. Mount in balsam.

EXAMPLE 4 | Preparation of a Wholemount of a Liver Fluke, Using the Carmalum Stain of Mayer

Although many people will be forced to rely on a supply house for their material, much better preparations can be made if the living flukes are obtained from a slaughterhouse. In this case, the flukes should be removed from the liver, where many will be found crawling upon the surface if the animal has been dead for some time, to a thermos flask containing physiological saline solution at a temperature of about 35°C, to which has been added a small quantity (approximately 1 g per liter) of gelatin. Flukes can be transported alive for relatively long distances in this solution, and every possible effort should be made to keep them alive until they have been brought to the laboratory and are ready to be fixed. In the laboratory the contents of the thermos flask should be poured into a dish and the worms transferred individually to another large dish containing warm physiological saline, where the last of the blood will be washed from them. Better preparations will be secured if time is taken to anesthetize the worms before fixing them, since most of the thick and opaque mounts like that shown in Fig. E4-2 result from an endeavor to fix an unanesthetized worm that has contracted during the course of fixation. Liver flukes are easy to anesthetize; the simplest method is to sprinkle a few crystals of menthol on the surface of the warm saline solution and leave the flukes in this for about ½ hr. Care should be taken that the worms do not die in this solution, and the process must be terminated when their motions are exceedingly slow and consist only of occasional feeble contractions rather than the active movements in which they were indulging when removed from the liver.

While the worms are being anesthetized, preparations for fixing them should be made. Take two sheets of ¼-in. plate glass, large enough so that the number of worms that are to be fixed can be laid on them, and place

Preparation of a Wholemount of a Liver Fluke, Using the Carmalum Stain of Mayer

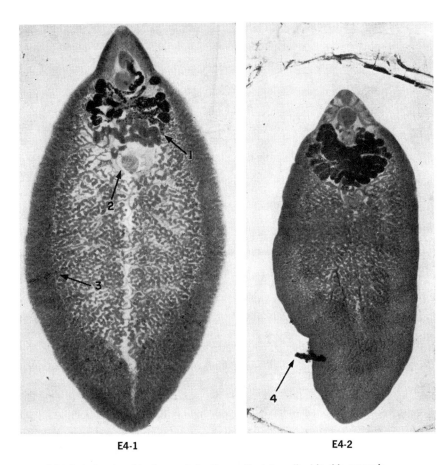

E4-1   E4-2

Figs. E4-1 and E4-2 | **Good and bad mounts by the method described in this example.**
Fig. E4-1. An excellent wholemount of a liver fluke. Notice at 1 that at least part of the uterus has been properly cleared and that at 2 there is good detail in the genital region. There is also at 3 a clear demarcation between the testes and yolk glands. Fig. E4-2. A thoroughly bad mount. All the good features shown in Fig. E4-1 are lacking. In addition, the specimen is contracted and distorted, indicating imperfect narcotization and imperfect flattening. There is also a large lump of dirt at 4.

upon the lower plate two or three thicknesses of a rather coarse filter paper or paper toweling. Blotting paper is too soft to be used for the purpose, and a good filter paper is much to be preferred to a paper towel. The selection of fixative must rest, of course, in the hands of the operator, but the author's preference is for the mercuric-acetic-nitric mixture of Gilson. This has all the advantages in sharpness of definition given by mercuric fixatives, and the addition of nitric acid appears to render the flattened worms less brittle in subsequent handling. Whatever fixative is selected, the sheet of filter paper is saturated thoroughly with it, and the anesthetized worms are removed from the physiological saline and laid out one by one about 1 in. apart on the paper. This must be done as rapidly as possible to prevent fixation from taking place before an additional layer of paper is placed on top and is saturated with fixative and the second sheet of glass placed on top of this. Assuming that the sheets of glass employed are of the size of a sheet of typewriter paper, it is suggested that about a 2-lb weight be placed on the upper sheet of glass. The whole should now be left for at least 12 hr before removing the glass and upper paper, picking the worms up one by one on a glass section lifter (metal cannot be used because of the presence of mercuric chloride), and transferring them to a large jar of fixative, where they may remain for another day to another week at the discretion of the technician.

At the conclusion of fixation the worms should be washed in running water for at least 24 hr. It has not been the author's experience that this fixative, followed by such washing, requires the use of iodine for the final removal of the mercury. At this stage, moreover, iodine tends to render the worms brittle, and the author would strongly recommend its omission. After thoroughly washing in water, the worms may be stained; the formula selected for the purpose of this example is the well-known carmalum of Mayer. Objects of this type are better stained by the additive process than by a process of differentiation; that is, they are better placed in an exceedingly weak solution and allowed to absorb the stain slowly than placed in a strong solution that requires subsequent differentiation. The best diluent for the stain is a 5 per cent solution of potassium alum. The extent of the dilution is dependent upon the choice of the operator and the size of the object to be stained. In the present instance a dilution of about 1 part of the stain to 100 parts of 5 per cent potassium alum would be correct. It is far more dangerous to have the solution too strong than it is to have it too weak, and, since it is an excellent preservative, the worms can remain in it for an indefinite period. The worms are merely placed in this diluted stain and left there until their internal structures have become clearly visible. It is suggested that they be examined at the end of a week, and subsequently every three days, until examination with a low-power binocular microscope, using a bright light from beneath, shows the testes to be darkly stained. At this point the worms are removed to a fresh clean solution of 5 per cent potassium alum and rinsed for a short time to remove all the adherent color. However, they will still be pink on the outside. Since the purpose of the stain is to demonstrate the internal organs, it is obviously desirable to bleach this outer layer in order to produce bright scarlet internal organs against a white background. In the experience of the author this may be done most readily with the aid of a potassium permanganate–oxalic acid bleach in the following manner: Prepare a solution of potassium permanganate so weak that it appears only a very faint pink. This is best done by adding a few drops of a strong solution to a beaker of distilled water. Then each worm is dropped individually into the solution and allowed to remain there until it has turned a bronzy brown on the outside. This appearance can best be detected in reflected light. Just as soon as the first bronze sheen appears on the outside of the worm it must be removed to fresh distilled water, where it can remain until all the other worms in the batch have been treated similarly. It will be necessary to renew the potassium permanganate from time to time, which is done by adding a few more drops of the stock solution to the beaker. The strength of the oxalic acid used to bleach the worm is quite immaterial; 2 or 3 per cent, arrived at by guesswork rather than by weighing, is an adequate solution. Since the bleaching of the surface is not nearly so critical as the deposition of the potassium permanganate, all the worms may be bleached at the same time by pouring off the distilled water and substituting the oxalic acid for it in the beaker. One or two twists of the wrist to rotate the worms in the beaker will result in their turning from a bronze sheen to a dead white. The oxalic acid is poured off without any waste of time, and the worms are washed in running tap water for an hour or so before being dehydrated in the ordinary way, cleared, and mounted in balsam. Little trouble will be experienced with curling of the worms if they have been fixed and treated as described. If they curl, they should undergo the final dehydration in 96 per cent alcohol while being pressed loosely between two sheets of glass, which will be sufficient to hold them flat.

SUMMARY

1. Anesthetize flukes in warm saline solution with menthol.
2. Place a layer of filter paper saturated with Gilson's fluid on a glass plate. Have ready another sheet of fixative-saturated paper and another glass plate.
3. Lay worms about 1 in. apart on paper on glass, cover with second sheet saturated with fixative, lay second glass on top, and add sufficient weight to flatten but not squash worms.
4. After about 12 hr remove flattened worms to jar of fixative. Fix 1 to 6 days. Wash in running water overnight.
5. Transfer to 1 part Mayer's carmalum diluted with 100 parts 5 per cent potassium alum. Leave until internal structures are stained clearly (1 to 3 weeks).
6. Wash in 5 per cent potassium alum.
7. Bleach surface with permanganate–oxalic acid.
8. Wash thoroughly, dehydrate, clear, and mount.

EXAMPLE 5 | Smear Preparation of Monocystis from the Seminal Vesicle of an Earthworm

Very few Sporozoa are available for class demonstration purposes, and the choice is practically limited to the inhabitants of the intestines of a cockroach or to the specimen under discussion.

The advantage of using Monocystis is that all the forms from the sporozoite to the trophozoite occur in the seminal vesicle of the earthworm and, therefore, may be made available on a single smear. The degree of infection among earthworms varies greatly, but it has been the author's experience that the larger the earthworm, the more likely the chance of a heavy infestation. At the same time, there is no use making a great number of smears for class demonstration purposes until a preliminary survey of a single smear has shown that the material will be satisfactory.

There is no need to kill or anesthetize the earthworm, which is simply pinned down in a dissecting tray and slit from the anterior end to about the sixteenth or seventeenth segment. The edges of this slit are pulled back and pinned into place, disclosing the large white seminal vesicles.

There should be available, before making the smear, a petri dish in which are a couple of short lengths of glass rod, a supply of the fixative selected, an adequate supply of clean glass slides, an eyedropper type of pipette, some 0.8 per cent sodium chloride, and some coplin jars of distilled water. Enough fixative is poured into the petri dish so that when a slide is laid on the pair of glass rods, its lower but *not* its upper surface will be in contact with the fixative. This level should be established with a plain glass slide before the smears are made.

Now slit the seminal vesicle of the earthworm and remove a drop of the contained fluid with the eyedropper pipette. This pipette is used to smear a relatively thick layer of the material on the center of one of the clean slides. Before it has time to dry, lay the slide face down in the fixative for a period of about 2 min. Next rinse the slide under the tap and examine it under a high power of the microscope after a coverslip has been placed over the smear. It is rather difficult to see the trophozoite stages in an unstained preparation, but no difficulty will be experienced in picking out the spore cases (pseudonavicellae) because of their relatively high index of refraction. It may be assumed that adequate numbers of the parasites are present if not less than three or four of these spore cases occur within the field of a $\times 10$ objective in a thick smear of this nature.

If the worm shows satisfactory infection, the remainder of the material from the eyedropper pipette is placed in a watch glass and diluted with 0.8 per cent sodium chloride until it forms a dispersion about intermediate in thickness between cream and milk. Working as rapidly as possible, as many smears as are required are made from this dilution. The dilution in question will not retain the parasites in good condition for more than about 5 min, but, if insufficient smears have been made in this time, it is easy to take a fresh supply of the seminal fluid from another vesicle and to dilute it in the fresh watch glass. The smears should be made with two slides in the manner described in Chapter 11, and each slide should be placed face downward in fixative for 3 or 4 min before being removed to a coplin jar of distilled water.

After they have been washed in water, the smears should be transferred to 70 per cent alcohol, where they can remain until they are ready for staining. Any stain may be used, but it is conventional to employ a hemotoxylin mixture. The method of staining is easy. The solution is made in accordance with the directions given and diluted to the extent of about 2 per cent with distilled water. The slides are placed in this diluted solution and left until examination under the low power of the microscope shows them to have been adequately stained and differentiated. Then they are rinsed exceedingly briefly in distilled water and dried. There is not the slightest necessity to employ any dehydrating agent, such as alcohol, for the smears should be sufficiently thin and the objects in them sufficiently well fixed that drying will not make the least difference. To complete the mount, a drop of the mountant selected is placed in the center of each smear, a coverslip added, and the whole put on a warm plate until it is dry.

SUMMARY

1. Make a temporary thick smear from the seminal vesicle of an earthworm. If plenty of spore cases are present, proceed with step 2; if not, test other worms until a heavily infected specimen is found.

2. Place two glass rods in a petri dish and lay a slide across them. Pour in just enough fixative to wet the lower surface of the slide.
3. Dilute the contents of the infected seminal vesicle with 0.8 per cent sodium chloride to the consistency of a thin cream.
4. Make a smear of cream as described in Chapter 11.
5. Lay each smear face down on glass rods in fixative for 3 to 4 min. Wash and accumulate in a coplin jar of 70 per cent alcohol.
6. Stain and differentiate smears. Dry in air. Mount in balsam.

EXAMPLE 6 | Smear Preparation of Human Blood Stained by the Method of Wright

An illustrated account of the preparation of a blood smear has been given in Chapter 10, and it is unnecessary to repeat the details here. Clean slides are an absolute necessity—Fig. E6-2 shows a smear made on a greasy slide—and an adequate number should be placed at hand before the blood is taken. There should also be available a drop bottle of Wright's stain and another either of distilled water or of a phosphate buffer at pH 6.4. Most supply houses that handle prepared Wright's stain offer a special buffer for use with it.

Human blood for smears is commonly drawn from a puncture wound in the ball of the forefinger, which should be compressed until the tip is suffused. The wound should be made either with a sterile hypodermic needle or with a special device manufactured for the purpose. A good drop of blood is then squeezed out and touched to the slide about 1 in. from the end (Fig. 11-1). A second slide (Figs. 11-2 and 11-3) is then used to push the blood into a thin smear, which is waved in the air to dry. A glance at the dried film through a $\times 10$ objective should show an erythrocyte distribution similar to that shown in Fig. E6-1. Smears thicker than this, which are used only for malaria diagnosis, require special staining methods, and thinner smears are unlikely to have an adequate distribution of leukocytes.

One of the dried smears is now taken and flooded with Wright's stain. For this purpose it may conveniently be rested across the open mouth of a coplin jar, although special racks are available if many slides are to be made at once. After 1 min, water or buffer is added to the stain on the slide in the proportion of 2 parts added fluid to 1 part stain. After 2 min the diluted stain is washed off with a jet of distilled water and the slide allowed to dry.

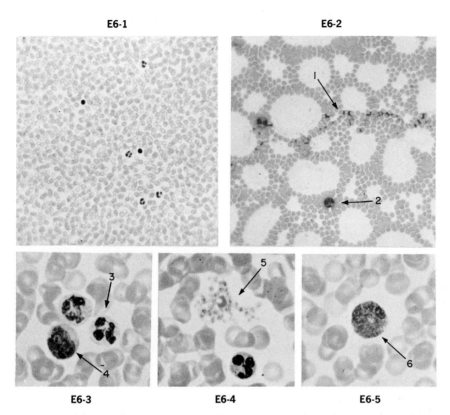

Figs. E6-1, E6-2, E6-3, E6-4, and E6-5 | Good and bad blood films by the method described in this example. Fig. E6-1 shows an excellent smear under low power. Compare with Fig. E6-2, in which grease has caused the overly thick film (notice overlapped corpuscles of 1) to coagulate. At 2 there is an imperfectly stained leukocyte partly masked by erythrocytes. Figs. E6-3, E6-4, and E6-5 show details of excellently stained leukocytes. Notice the polymorph at 3, the monocyte at 4, the platelets at 5, and the exceptionally well-demonstrated eosinophil at 6.

The success of the stain is judged by an examination (Figs. E6-3 to E6-5) of the leukocytes under a ×90 immersion lens. The nuclei of polymorphs should be clear dark blue, the platelets sharply distinguished, and the granules of eosinophils bright red. Inadequately stained nuclei usually indicate too short a period of exposure to undiluted stain. Poorly stained eosinophils are generally the result of adding too much water or buffer. Batches of Wright's stain vary widely in their performance, and a few trials are necessary before uniformly successful results can be obtained.

Smears that appear to contain no leukocytes are the result of using too large a drop of blood. Capillary attraction draws the large white cells to the edges of the pushing slide, and they will be found embedded in the thick strips of blood that outline the periphery of smears so made.

Most blood smears are regarded as temporary mounts made for diagnostic purposes. They may be stored dry for about a year, but the immersion oil must be washed off with benzene each time they are examined, to avoid the collection of dust. Smears that are to be permanently preserved should be mounted in a neutral mountant, like that of Mohr and Wehrle (Chapter 9), under a No. 0 coverslip.

SUMMARY

1. Make smear (Chapter 11; Figs. 11-1 to 11-3).
2. Flood with Wright's stain.
3. Add water or buffer to stain on slide.
4. Rinse.
5. Preserve dry or mount in Mohr and Wehrle's medium.

EXAMPLE 7 | Staining a Bacterial Film with Crystal Violet by the Technique of Lillie

This technique is so simple that it would be scarcely worth the trouble to describe it were it not necessary for the benefit of those who have never previously handled bacterial material and who may wish to attempt this for the first time.

The only tools and reagents necessary are a clean glass slide, a wire loop of the type used normally in bacteriology, a drop bottle containing the crystal violet stain, and a wash bottle containing distilled water.

The freshly flamed wire loop is touched as lightly as possible either to the surface of the medium in a test tube or to the surface of the colony in an agar petri-dish culture, and it is then touched lightly to the center of the clean slide to transfer the bacteria. The only mistakes likely to be made by the beginner are having too great a quantity of material and having it on a greasy slide. The result of this combination is seen in Fig. E7-1. It must be remembered that the specimen is to be examined under an oil-immersion lens, so that the smallest possible smear, derived just by touching the slide with the moist platinum loop, will have an ample area for the purpose required.

If the microorganisms have been taken from a liquid culture that has not yet reached a very thick stage of growth, the spot on the slide now may be permitted to dry in air; but if the bacteria have been taken from a colony on the surface of a dish, it is necessary to dilute the drop on the slide. For this purpose, a small quantity of water in the same platinum loop is touched to the same spot and rubbed backward or forward once or twice, and then the slide is permitted to dry.

As soon as the slide has dried, which should be only a moment or two if a sufficiently small quantity of suspension of the specimen has been used, it is taken and heat-fixed in the flame of a bunsen burner or a spirit lamp. This is done by passing the slide rather rapidly twice

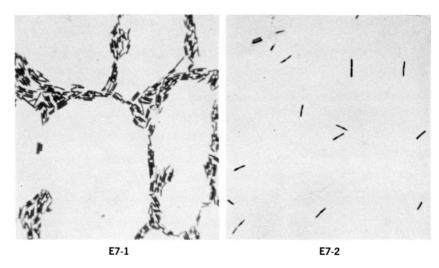

Figs. E7-1 and E7-2 | **Bad and good bacterial smears.** Fig. E7-2 shows properly distributed bacteria in the smear. Fig. E7-1 is the result of applying too heavy a smear on a dirty slide.

through the flame. The actual temperature should not exceed about 80°C, and it is customary to hold the slide smear downward as it passes through the flame. Care must be taken that the slide is dried before it is thus quickly flamed, or, of course, the bacteria will burst and be worthless.

On the flamed and dried smear is placed a drop of the selected stain, leaving this in place for about 30 sec. The time is not critical, and any time between ½ and 1 min is perfectly satisfactory. It will be noticed that the stain frequently evaporates slightly, leaving a greenish film on the surface, and it is, therefore, better to wash it off with a jet from a wash bottle than to rinse it off. This jet should be directed from the fine orifice of the wash bottle at an angle of about 30° to the slide and should be intended to hit the edge of the drop. The jet will instantly lift off and float away the surface film as well as wash excess stain out of the preparation. The preparation is now permitted to dry. When dry, it may be touched with immersion oil and examined under the oil-immersion lens. If it is desired to maintain this specimen permanently, it may also have a drop of Canada balsam placed on top of it and a very thin coverslip pressed into position. These specimens, however, are so easy to prepare that it is rarely necessary to preserve them permanently. The dried smear, if kept free from dust, may be preserved for as long as 1 year without further preparation.

SUMMARY

1. Use a wire loop to make a small thin smear of a diluted bacterial culture on a clean slide.
2. Dry smear in air; then pass it two or three times through a flame.
3. Place a drop of crystal violet stain on the smear, allow it to remain for 30 sec, and wash off with a jet of water from a wash bottle.
4. Dry smear without heat.

EXAMPLE 8 | Demonstration of Gram-positive Bacteria in Smear Preparation by the Method of Gram

For the benefit of those who are not acquainted with bacteriology, it may be said by way of introduction that it has been customary since the time of Gram to utilize the reactions of bacteria to iodine mordanting as a basis of diagnostic classification. All bacteria, without reference to their nature, may be stained by the method given in Example 7, but there are some bacteria from which this stain can be removed by the action of an iodine–potassium iodide solution reinforced with alcohol. The bacteria from which the stain is not removed are known as *Gram positive;* those from which the stain is removed are known as *Gram negative.* The solutions required are the violet used in Example 7 and Gram's iodine solution, the formula for which is given in Chapter 7. Iodine is soluble in strong solutions of potassium iodide but is only slightly soluble in weak solutions. If the total quantities of iodine and potassium iodide are placed in the total quantity of water, a period of as long as a week may elapse before a solution is complete; if the dry iodine and the dry potassium iodide are mixed together and a few drops of water added, the whole will go almost instantly into solution.

Presuming that one is working with a pure culture of bacteria, a smear is prepared as described in Example 7, taking the same precautions as to dilution mentioned there if the material is obtained from a bacterial colony. The smear is dried and flamed, and a drop of crystal violet is poured on it from a drop bottle exactly as in the previous example. In this instance, however, it is not desirable to extract too much of the stain with water. After the stain has been acting for 2 min or so, it is not removed, as in Example 7, with a jet from a wash bottle, but the entire slide is rinsed rapidly in a fairly large volume of water and a drop or two of the iodine solution poured over it. If many slides are being stained, it is probably simpler to drop each slide into a coplin jar containing the iodine solution than to pour iodine on it. After the iodine has been permitted to act for a period of 1 min, the whole slide is given a quick rinse to remove the excess iodine and then placed into absolute alcohol until no more color comes away; unless the film is very thick, this will appear to decolorize it completely. It is passed from alcohol to water, which instantly stops differentiation, and then dried. Varying types of bacteria require varying periods of differentiation, but it is best for the beginner to use absolute alcohol until no more color comes away rather than to endeavor to control the differentiation under the microscope.

Though this is a Gram preparation by the original technique, it is customary nowadays to provide a counterstain of a contrasting color to bring clearly into evidence any Gram-negative organisms that may be mixed with Gram positive. A 1 per cent solution of safranin is widely employed, and in this case, the second red contrasting stain is allowed to act from 5 to 10 sec and is then washed off with water, and the slide is then dried and examined under an oil-immersion objective.

SUMMARY

1. Follow steps 1 to 3 of Example 7 but allow stain to act for 2 min. Then wash it off under tap.
2. Treat stained smear with Gram's iodine for 1 min. Rinse under tap and soak in absolute alcohol until no more color comes away.
3. If desired, counterstain 5 to 10 sec in 1 per cent safranin and wash it off with water before drying.

EXAMPLE 9 | Demonstration of Tubercle Bacilli in Sputum by the Technique of Neelsen

When Neelsen published his original technique for the demonstration of tubercle bacilli, the standard magenta solution used for the staining of bacteria was that proposed in the previous year by Ziehl. As a result, Neelsen's technique was referred to as a modification of Ziehl's, and to this day the hyphenated term Ziehl-Neelsen is applied to almost any method for the demonstration of tubercle bacilli in sputum regardless of the author of the technique.

It is proposed here to describe the original technique of Neelsen, leaving it to the technician to determine which of the many other methods given in Gray's "Microtomist's Formulary and Guide" can be more readily applicable to his problem. It may be said in favor of the technique of Neelsen that it gives a far better differentiation of tubercle bacilli than some of the more recent methods which, although they give good preparations, tend to cause certain errors of diagnosis through the ability of other bacteria to withstand the lower concentration of acids employed.

It is to be presumed that the sputum collected from the patient will have been placed at the disposal of the technician in the glass vessel in which it was secured. It should be examined to see whether or not any small yellowish particles exist in it, and, if they do, one of these particles should be extracted carefully with a sterile bacteriological wire loop and utilized for the preparation. Even if no such particles are visible to the naked eye, it is possible that tubercle bacilli will be present, but due consideration should be given to some method of concentrating these bacilli before making the smear. The standard method of concentration is to hydrolyze the sputum to the extent necessary with the aid of a weak solution of potassium hypochlorite, which is known to be without action on tubercle bacilli. For a long time a proprietary compound known as *Antiformin,* which is a strongly alkaline solution of sodium hypochlorite, has also been employed for the same purpose. About equal quantities of the selected solution and the sputum are placed in a centrifuge tube. The tube is incubated in a serological water bath for about 10 min and then centrifuged rapidly, the smear being made from the denser portions that remain at the bottom of the tube.

Whichever method is employed, the quantity of material removed by the sterile loop should be about the size of a large pinhead. That is, a great deal more material should be employed than in the preparation of a simple bacterial smear from a known culture for the reason that large areas have to be searched in the interests of diagnostic accuracy. This pinhead of material must be spread over the largest possible area of the slide, which is most readily accomplished by pressing another slide on the first and then drawing the two slides apart with a lateral motion. Both slides are dried in air and flamed as has been described in previous discussions of bacteriological preparations.

The solutions required for the original Neelsen technique are the carbolmagenta solution of Ziehl, a 25 per cent solution of nitric acid, and a 1 per cent methylene blue solution. The slide is first flooded with a large quantity of the magenta solution and then placed on a metal sheet, where it should be warmed underneath by a bunsen flame to the point where the stain is steaming briskly but no bubbles have appeared. If it shows signs of drying, fresh quantities of the magenta solution should be added to it. Either it may be left at this temperature for 3 to 5 min or, as is customary in modern practice, it may be raised to steaming, permitted to cool, again raised to steaming, permitted to recool, and so on, until four such cycles have been completed. The slide is washed in a large volume of tap water until no further magenta comes away and is then placed in 25 per cent nitric acid until it is almost completely decolorized. It cannot be decolorized too far, but usually there will be found, even after prolonged exposure to the acid, a faint pink coloration of the background. The slide is next washed in running water until all the acid is removed. Finally, it is treated with a blue stain for about 2 min to provide a contrasting coloration of any other bacteria present and is washed thoroughly, dried, and examined in the customary manner.

It must be emphasized that this technique as described is designed specifically to show tubercle bacilli and is so violent that many bacteria that are acid-fast to less strong acids will be decolorized.

SUMMARY

1. Smear yellow flecks or centrifuged concentrate from sputum between two slides. Separate, dry, flame, and stain both sides.
2. Flood smear with Ziehl's magenta, warm to steaming, and cool. Repeat warming and cooling three times.
3. Wash under tap and decolorize in 25 per cent nitric acid.
4. Wash off acid and counterstain, if desired, in 1 per cent methylene blue for 2 min.
5. Wash in water and dry.

EXAMPLE 10 | Preparation of a Squash of the Salivary Glands of Drosophila Stained in LaCour's Acetic Orcein

The technique described in this example may be applied to any squash material of either plant or animal origin. Drosophila has been selected because it is usually available as a surplus commodity in any laboratory in which genetics is taught. Before commencing, it is necessary to set out a supply of LaCour's acetic orcein, a pair of sharp needles, a number of clean slides and coverslips, and a supply of cold-blooded Ringer's solution. This has the composition:

Sodium chloride	0.65 g
Potassium chloride	0.025 g
Calcium chloride	0.03 g
Water	100 ml

It is also necessary to have available a supply of bibulous paper of the type sold by scientific supply houses for blotting slides. Neither commercial blotting paper nor laboratory filter paper is as good for this purpose.

Good preparations of the salivary gland chromosomes of Drosophila can be made only from the third instar larva. These may readily be distinguished by their large size and sluggish movement, which precedes pupation. They are usually to be found crawling up the glass side of the cream bottles in which laboratory Drosophila are customarily cultivated.

Experts in this technique usually dissect the salivary gland from the larva directly in the stain, but it is better for beginners to do this dissection in cold-blooded Ringer's solution.

Lay out a clean slide and place a large drop of stain on it. Pick out a number of larvae of the correct stage and lay them on one side of another slide. Lay this slide on the stage of a binocular dissecting micro-

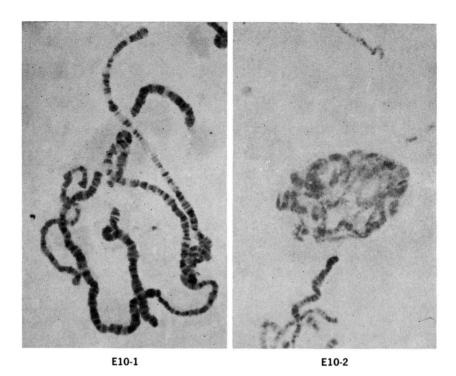

E10-1   E10-2

Figs. E10-1 and E10-2 | Good and bad salivary-gland chromosomes prepared by the method described in this example. Fig. E10-1 shows well-spread chromosomes with the bands clearly marked. Fig. E10-2 is the result either of letting the salivary gland dry before applying the stain or of imperfect crushing. The dark background in both pictures is caused by the presence of residual stain and cannot be avoided in preparations of this type.

scope against a black background. Place a drop of Ringer's solution on one end of this slide and transfer a larva of appropriate age to a point about ⅛ in. to the left of the drop. Hold the body of the larva firmly with a needle in the left hand and insert the point of the right-hand needle into the center of the mouth parts. With a continuous steady pull, drag the mouth parts, and whatever comes with them, into the drop of Ringer's solution. The salivary glands will now be clearly apparent as two translucent sausage-shaped bodies on each side of the esophagus. Both needles are now used to detach the mouth parts, and as much adherent fat as possible, from the glands. The glands are now pushed onto the point of the right-hand needle, lifted out, and immediately placed in the drop of stain on the second slide.

As many glands as required are accumulated in the drop of stain and left to soak for at least 5 min before the coverslip is pressed on.

After the glands have soaked long enough, a large (⅞-in.-square) coverslip is pressed on the drop. A sheet of bibulous paper is then pressed on top and the ball of the forefinger rocked backward and forward with a light pressure over the center of the coverslip to force out the excess stain. A clean piece of paper is now pressed over the coverslip and again pressed with the ball of the finger, this time with sufficient pressure completely to flatten the glands. This blotting and pressing should be repeated several times until only a faint pink color remains under the coverslip. Examination under the high power of the microscope will now show the chromosomes properly separated with the bands sharply stained in the manner illustrated in Fig. E10-1. Unsuccessful preparations of the type shown in Fig. E10-2 are usually the result of permitting the gland to become partially dried before it is stained.

Temporary preparations by this method continue to improve in both clarity and density of staining for at least 24 hr and, if sealed with molten paraffin, may be kept for some weeks.

If it is desired to make these slides permanent, it is necessary to secure a supply of dry ice, a safety-razor blade, and coplin jars containing 95 per cent alcohol, absolute alcohol, and xylol. Lay each slide, coverslip up, on a block of dry ice and let it remain there for at least 5 min. It is possible to freeze slides on the block holder of a freezing microtome, but this method is less certain.

When the slide has been adequately frozen, place a needle against the left-hand edge of the coverslip and push the edge of the safety-razor blade at a low angle against the right-hand side of the coverslip, which can, by this method, be removed without disturbing the chromosomes. Throw the coverslip away and transfer the slide immediately to a coplin jar of 95 per cent alcohol for about 2 min. Then transfer to absolute alcohol for a similar length of time before placing it in xylol. When cleared by the xylol, a drop of balsam may be added and a coverslip placed on top in the usual manner.

SUMMARY

1. Select third larval instars of Drosophila and place them on a clean slide.
2. Place a drop of LaCour's acetic orcein in the center of another clean slide.
3. Place a drop of Ringer's solution on the end of the slide on which are the larvae. Push a larva near the drop. Pull the salivary glands into the drop. Trim them.
4. Transfer the salivary glands to the drop of stain. Leave 5 min.
5. Place a ⅞-in. coverslip on drop. Cover with a piece of bibulous paper and press firmly. Repeat with a second piece of bibulous paper.
6. A successful preparation may be made permanent by freezing on dry ice before removing the coverslip to 70 per cent alcohol. Then dehydrate and mount in balsam.

EXAMPLE 11 | Preparation of a Transverse Section of a Root, Using the Acid Fuchsin—Iodine Green Technique of Chamberlain

This is the simplest of all the preparations described in this part of the text and can be recommended unhesitatingly to the beginner who has never previously prepared a section of any type. This preparation is designed to show only the skeletal outlines of the cells and is not intended to demonstrate in any way their cytological contents, which are removed in the course of preparation. If cytological detail in a botanical section is required, a paraffin section should be prepared and stained by the method given in Example 13.

It does not matter from what source the root is obtained, but it is recommended that the beginner select some soft root of about ⅛ in. or somewhat less in diameter. If the root is collected from a living plant, it should be washed thoroughly to remove any adherent sand grains, which will spoil the edge of the cutting knife, and then preserved in 96 per cent alcohol until required. The 96 per cent alcohol should be changed as it becomes discolored, but with this precaution the specimens may be preserved indefinitely.

It is even possible to make preparations of this type from dried roots that have been preserved in a herbarium. The best reagent for swelling and softening these dried preparations is a 10 per cent solution of phenol in lactic acid. The lactic acid employed is the ordinary commercial solution in which the phenol should be dissolved immediately before it is required. Pieces of the dried root are placed in a reasonably large volume of this material and heated over a low flame to a temperature of about 50°C. Within 10 or 15 min a completely dried herbarium specimen will have become swollen out to its normal size and softened to the extent that sections may be cut readily from it.

The method of sectioning selected does not particularly matter. Since the sections cannot in any case be subjected to the first process while they are attached to the slide, there is no real advantage in embedding in paraffin and cutting in this medium if an ordinary hand microtome is available from which sections can be taken by the conventional method (see Chapter 12) of holding them either in pieces of pith or between the cut halves of a carrot. If the sections are cut by hand, they may be transferred immediately after they are cut to a dish of 20 per cent alcohol and from there to water; if they are cut in paraffin, they should be at least 20 μ thick. The ribbon, as it is removed from the microtome, should be dropped directly into a watch glass of xylene, in which the paraffin will dissolve readily. The individual sections are then removed from the xylene with a section lifter, passed through absolute alcohol for the removal of the xylene, and then graded down through alcohols to water. By whatever method the sections are produced, they are accumulated in a small dish of distilled water. These sections, of course, will retain the cell contents, which must be removed in order that the sections may be turned into true skeletons.

The best reagent to use for skeletonizing sections of plant tissue is either potassium or sodium hypochlorite. Ordinary bleaching solutions sold for household purposes under various trade names are not suitable since they contain large quantities of calcium hypochlorite. If, however, the pure salts are not available, the household solution may be employed by adding to it enough of a solution of potassium or sodium carbonate to precipitate the calcareous contents and by filtering off the solution before use. If the pure salts are available, a 1 per cent solution may be employed conveniently. The sections are removed from the distilled water on a section lifter and transferred to a watch glass of the sodium or potassium hypochlorite solution. If the sections are made from material that has been preserved in alcohol, this solution should be used cold, but usually it must be warmed if it is to have the desired effect on materials that have been resurrected from a dried condition. In either case the operation should be watched very carefully under a low power of the microscope, and the action of the hypochlorite should be discontinued as soon as the cells are seen to be free of their contents. If the mounter is completely inexperienced in this field and is unable to determine the point at which the operation should be stopped, it is recommended that a single section be taken and the skeletonizing followed under a microscope while it is timed. It will be seen that when the operation has gone too far, the finer cell walls present will be dissolved by the solution. If the time required for the first of the cell walls to dissolve is carefully recorded and one-half of this time taken for the subsequent sections, these will be cleaned perfectly without the slightest risk of damage to their walls. After they are removed from the hypochlorite solution, the sections should be washed thoroughly in several changes of distilled water and then passed into a solution of 1 per cent acetic acid, where they are rinsed several times and then rewashed in ordinary water until the wash water no longer smells of the acetic acid. The skeletonized sections, from as many roots as it is desired to cut at one time, should be accumulated in water until one is ready to stain them, or they may be preserved indefinitely in 90 per cent alcohol.

The stain recommended in the present case is freshly prepared for use by mixing equal parts of a 0.2 per cent acid fuchsin solution and a 0.2 per cent iodine green solution. The mixed stains do not remain usable for much longer than 1 day, but the two separate stock solutions may be kept for an indefinite period. The differentiating solution, which is 1 per cent acetic acid in absolute alcohol containing 0.1 per cent iodine, is also stable. The staining solution should be placed in a small corked vial or stoppered bottle and the sections transferred to it from the water. They should remain in stain for about 24 hr. It is recommended that they not be left longer than 36 hr because they may suffer from a precipitate over the surface, and when the staining period is concluded the contents of the vial should be tipped out into a large watch glass. It will usually be found that some of the sections remain behind, stuck to the side of the vial from which they are removed. Under no circumstances should the vial be rinsed with anything except the staining solution, which should be poured back from the watch glass (the sections will have settled to the bottom), swirled around, and returned to the watch glass. In this way in a short time all the sections may be removed from the vial to the watch glass. In a second watch glass or even a small crystallizing dish is placed an adequate quantity of the differentiating solution. Each section is removed individually with a section lifter from the stain and placed into the differentiating solution, where it may be watched under the low power of a microscope as the dish is rocked gently from side to side. Differentiation will usually take place within 2 or 3 min and may be determined without the least difficulty when the lignified tissues are found to be of a bright clear green, leaving a bright red in the nonlignified tissues. This process of differentiation is also one of dehydration, so that the sections may now be removed with a section lifter from the differentiating solution and placed in a clearing agent. The author's preference is for terpineol, which has all the advantages of clove oil without the disadvantage of making the sections brittle so that they crack on mounting. All the sections may be passed through the differentiating solution and accumulated in terpineol. They may remain in terpineol until removed to a slide, where they are covered with balsam and mounted in the normal manner.

Sections prepared in this manner are permanent, and the process is so simple that it can be recommended most warmly as an introduction to plant-section-staining techniques for an elementary class. The sections, however, are differentiated clearly enough to be used for instructing a class at any level, and generally they will be found much better for this purpose than the complex quadruple-stained sections in which the cytological detail all too often tends to obscure the clarity of morphological detail, the chief requirement in this type of teaching.

SUMMARY

1. Cut hand sections of plant material hardened and preserved in alcohol. Accumulate sections in water.
2. Place sections in 1 per cent sodium or potassium hypochlorite until cell contents have been dissolved, leaving clean bleached skeleton.
3. Wash in water, rinse in 1 per cent acetic acid, and wash again in water.
4. Stain 24 hr in equal parts 0.2 per cent fuchsin and 0.2 per cent iodine green.
5. Differentiate sections individually in absolute alcohol containing 1 per cent acetic acid and 0.1 per cent iodine.
6. Transfer directly to terpineol and mount in balsam.

EXAMPLE 12 | Preparation of a Transverse Section of the Small Intestine of the Frog Stained with Celestine Blue B—Eosin

This is the simplest example of paraffin sectioning that can be imagined, and it may well serve as an introduction to this type of technique either for a class or for an individual. The intestine of a frog has been selected, rather than of any other animal, because of the availability of this form in laboratories, but any small animal may be substituted in its place.

Before killing the frog, it is necessary to have on hand a selected fixative. Since this is intended to be an example of the utmost simplicity, it is suggested that the dichromate-mercuric-formaldehyde mixture of Helly (Chapter 6) be employed. This fixative is entirely foolproof because objects may remain in it for weeks without damage, and it also permits excellent afterstaining by almost any known technique. If only a piece of intestine is to be fixed, 100 ml of fixative will be sufficient. There is no reason why any other organ in the animal (with the exception of the central nervous system) should not be preserved in this fluid for subsequent investigation.

The frog is killed by any convenient method, but it is usually best for histological purposes to sever a large blood vessel and permit as much blood as possible to drain out from the heart before opening the abdominal cavity and removing the intestine. One or more lengths of about ⅓ in. should be cut from the intestine and then transferred directly to the fixative, where they may remain overnight.

When they are next required, the specimens should be removed from fixative, washed in running water for a few hours, and then transferred directly to 70 per cent alcohol. The easiest method of washing objects of this size in running water is to take one of the coplin jars previously described, fill it with water and insert the specimen, and then attach a

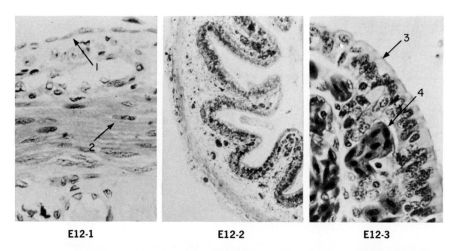

E12-1   E12-2   E12-3

Figs. E12-1, E12-2, and E12-3 | **Details from an excellent preparation produced by the method described in this example.** Fig. E12-2 shows the low-power view to indicate the relative staining of the various parts. Fig. E12-1 shows the edge of the section. Notice the undamaged mesentery at 1 and the clearly differentiated nuclei in the circular muscle at 2. Fig. E12-3 shows the edge of one of the villi. Note the smooth undamaged membrane at 3 and the erythrocytes at 4, in which the black nuclei are sharply differentiated in spite of the heavy eosin pickup.

cover of coarse cheesecloth with a rubber band. The jar is placed in the sink, and a narrow stream of water is permitted to fall on it from the tap. It will be found that the specimen will swirl around in the jar in a most satisfactory manner. This simple device saves all the trouble of rigging up glass tubes and boring corks to make the cumbersome apparatus sometimes recommended for the purpose.

The actual procedure of embedding has already been described in some detail. The specimen is transferred after 24 hr in 70 per cent alcohol to 96 per cent alcohol. It is better to use a large volume of this alcohol and to suspend the object in it than to use relatively small volumes frequently changed. It is recommended, therefore, that a wide-mouthed stoppered jar of about 500 ml capacity be obtained and a hook inserted into the center of its stopper, from which the object can be suspended. The majority of wide-mouthed glass-stoppered jars have a hollowed under surface which may be filled with plaster of paris, and a glass hook, which is easily bent from thin glass rod, may be inserted in the liquid plaster. Naturally this must be done some days beforehand, and the plaster must be thoroughly dried out in an oven before the jar is used for dehydrating. If the worker does not wish to go to this much trouble, it is also easy to screw a small metal pothook into the under surface of a plastic screw cover for a jar of the same size. Alcohol is so hygroscopic, however, that it is better to employ a glass-stoppered jar, the stopper being greased with stopcock grease or petrolatum for a permanent setup. An object as coarse as the one under discussion may be suspended in a loop of thread or cotton directly from the hook or, if this is not desirable, may be enclosed in a small fold of cheesecloth for suspension. After 24 hr in this volume of alcohol, the object will be completely penetrated by the 96 per cent alcohol and should be transferred to absolute alcohol, using the same volume in a jar of similar construction. It is useful to place about a ¼-in. layer of anhydrous copper sulfate at the bottom of the absolute alcohol jar, not only to make sure that the alcohol is absolute but also to indicate, as it changes to blue, when this jar should be removed from service. Of the many dealcoholizing (clearing) agents that may be used, the author would select benzene in the present case because it is less liable to harden the circular muscles of the intestine than xylene. Since benzene is lighter than absolute alcohol, it is not possible to use the hanging technique for

Figs. E12-4 and E12-5 | **Details from a thoroughly bad slide made by the method described in this example.** Fig. E12-4 shows one villus. Notice at 1 the tears caused by a blunt knife and the folds that have resulted from imperfect flattening of the section. Compare 3 and 4 in Fig. E12-5 with 3 in Fig. E12-3. This animal had evidently been dead for some time before the section was fixed, so that autolysis has destroyed much of the epithelium and so damaged the rest of the tissue that sharp differentiation is not possible.

E12-4

E12-5

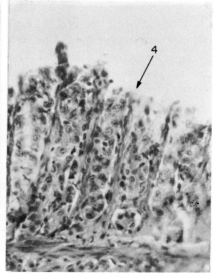

clearing, and the object should be placed in about 25 ml of benzene, which should be changed when diffusion currents no longer rise from the object. This is likely to be about 6 hr for an object of the size under discussion, and a second bath of at least 6 hr should be given.

It is necessary to select the medium in which embedding is to be done. The author would recommend the rubber paraffin of Hance, which must, of course, have been prepared some time before. The melting point of this medium is about $56°C$. The oven should contain three stender dishes (see Fig. 5-5) as well as a 500-ml beaker containing about a pound of the embedding medium. The object is removed from benzene, drained briefly on a piece of filter paper, and placed in one of the stender dishes, which has been filled to the brim with the molten embedding medium. Under no circumstances should a lid be placed on the stender dish because it is desirable that as much as possible of the benzene evaporate while the process of embedding is going on. After about 3 hr the specimen should be removed to fresh wax in the second stender dish, where it may remain about 2 hr, and then to the third stender dish, where it should not remain for more than 1 hr.

Shortly before the end of this last hour, a decision should be made as to what type of vessel is to be used for casting the block. It would be difficult to improve on a paper box for this object. When the box is made (it should be of ample size), it is moistened at the bottom and placed on a slab of glass in the manner described in Chapter 12. It should then be filled about halfway with embedding material from the beaker and allowed to remain until about half of this wax has congealed on the bottom. An object like the one under discussion is better handled with old forceps than with a pipette. The forceps should be warmed in a flame to well above the melting point of the wax and moved backward and forward across the surface to melt the solid film that has formed. Then the object is picked up rapidly from its stender dish and placed in the wax, and enough fresh wax from the beaker is added to make sure that there will be as much solid wax above, as there is underneath, the specimen. Blocks of this nature shrink greatly, and it will probably be best to fill the box entirely. As soon as the box has been filled, the forceps should be warmed again and passed backward and forward around the object to make sure that no film of unmolten wax, which would cause it to cut badly, remains. The wax in its box should now be blown on until it starts to congeal on the surface. Then it is picked up very carefully with the fingers and lowered into a dish of water at room temperature (the water does not quite reach the top of the box). If it is thrust under the surface at this point, all the molten wax will come out and the block will be useless. As soon as the block has congealed throughout, it is thrust under the surface of the water, and something is laid on it to keep it at the bottom. It should be left in the water for at least 5 or 6 hr or, much better, overnight.

The block should be trimmed so that there is at least as much wax on each side of the object as there is in the object itself. This amount of wax would be excessive for serial sections, but for the preparation of individual sections of this type in an example given for the benefit of the beginner, this quantity is desirable. When the microtome is set up and the knife is sharpened in the manner previously described, the block is mounted. This process has been described in some detail and need not be repeated here. The block having been trimmed to size and mounted, there remains only the actual cutting. The handle of the microtome should now be rotated rapidly and the beginnings of the sections observed. There is no need to worry if the section curls to one side or the other during this preliminary period since the entire thickness of the block will not be cut until 20 or 30 sections have been removed. As soon as the knife is seen to be approaching the object and the block in its entirety is being cut, the ribbon must be observed most carefully to see that it is suffering from none of the common defects indicated in Table 2. Should the ribbon not be coming perfectly, various suggestions given in Table 2 may be tried until a perfect ribbon is obtained. Since, in this case, a series of sections is not needed, it is unnecessary to cut a longer ribbon than one that will contain the actual number of sections required, with a few left over for emergencies. However, it is a great mistake to throw away partially cut blocks of this nature since they may be stored indefinitely, and one never knows when further sections may be required. The block should be labeled before being placed in its solution. This is easily done by writing the appropriate label on a piece of paper and fusing this into an unwanted portion of the block with a hot needle.

Each individual section is now cut from the ribbon and mounted on the slide in the way described in Chapter 12. Before actually using Mayer's albumen to mount the sections, it is necessary that the slides be cleaned. No two people have ever agreed as to what is the most desirable method of doing this. The author rubs the slide briskly with 1 per cent acetic acid in 70 per cent alcohol and dries it by waving it in the air. Another method of cleaning the slide, which yields equally good results, is given in Chapter 11. After the sections are cut, several drops of the diluted adhesive are placed in the center of each slide. One of the individual sections is taken up with the tip of a moistened brush and placed on the adhesive. As soon as the section has been placed on the fluid, the slide is lifted up and warmed carefully over a spirit lamp until the section is flat but the paraffin not melted, and then the superfluous liquid is removed carefully with the edge of a filter paper. The slide is put in a warm place to dry. If the drying period is to be prolonged, it is well to place a dust cover over the slide since grains of dust falling upon it will adhere just as tenaciously to the adhesive used as the specimen itself.

It is proposed in the present example to stain the slide in the simplest possible manner with celestine blue B followed by eosin. The formulas for these stains will be found in Chapter 7. Of course, any of the other dyes recommended there may be substituted.

When the section is perfectly dry, it is turned upside down, and light is reflected from it to see whether or not the section is adherent to the glass. If there is any air gap between the section and the glass, a brilliant mirror will be formed, so that, in a preparation as simple as this, the slide had better be thrown away. Those slides which are perfectly adherent are warmed over a flame until the wax is melted and then dropped into a jar of xylene, where they remain until a casual inspection shows the whole of the paraffin to have been removed. They are put into another jar of xylene for at least 5 min and then into a jar of equal parts xylene and alcohol for a further 5 min. This treatment is followed by 5 min in absolute alcohol and then by direct transference to distilled water. After the slides have been in distilled water for a few minutes, each slide should be lifted and inspected to make sure that the water is flowing uniformly over both the slide and section. If it tends to be repelled by the section or if a meniscus is formed around the section, this is evidence that the wax has not been sufficiently removed, and the slide must be transferred first to 96 per cent alcohol to remove the excess water, then to absolute alcohol until perfectly dehydrated, and finally through absolute alcohol to xylene, where it remains until the wax has been completely removed, before being brought down again as previously indicated. The slides may be taken down through xylene and alcohol one at a time and accumulated in distilled water until they are required. When all slides have been accumulated in distilled water, they are transferred to the celestine blue B staining solution for 1 min. One of the most useful features of this stain is that it is almost impossible to overstain in it. It is even possible to leave sections overnight without staining the cytoplasm to a degree that requires differentiation. Each slide is now dipped up and down in the eosin solution until a casual inspection shows the background to be yellowish pink. The staining of the background in a case like this is entirely a matter of choice; some people prefer a faint stain and others a darker stain, although it must be remembered in judging the color that the section will seem darker after it has been cleared than it does in water.

As soon as the time required to produce the desired degree of staining has been found by staining one slide, the remainder of the slides are placed in the eosin solution, left the appropriate time, and then transferred and left in distilled water until no more color comes away. The slides are passed from distilled water to 96 per cent alcohol, where they are left for about 5 min. Then they are passed to fresh 96 per cent alcohol, where they are left for 5 or 6 min before being passed to absolute alcohol. The purpose of using the 96 per cent alcohol is not to diminish diffusion currents but simply to save diluting the absolute alcohol by passing slides directly from water to it. After the slides have remained for 5 or 6 min in absolute alcohol, a single slide is passed into the absolute alcohol-xylene mixture for perhaps 2 min and then passed to xylene. On examination the slide should be transparent with only a faint opalescence. One of the commonest faults in mounting sections is dehydrating them imperfectly. If there is any water, which has been carried through the process, in the xylene (in which water is soluble to the extent of about 0.2 per cent), this water will be extracted by the section, which is in itself an excellent dehydrating agent. There is a world of difference between a perfectly cleared (that is, glass-clear) slide and one that is only more or less dehydrated so that it appears faintly cloudy. If the slide does not appear to be sufficiently dehydrated, all the remaining slides should be transferred to fresh absolute alcohol and another one tried. When it has become apparent from the examination of the test slide that dehydration is perfect, the remaining slides may be run up through absolute alcohol and xylene and accumulated in the final jar of xylene.

Each section must be mounted in a solution of dried balsam in xylene. All that is required for mounting is the cleaning of the appropriate number of coverslips, for which, in the present instance, a ¾-in. circle would be admirable. Again, individual preference may determine the manner of cleaning. The author cleans his coverslips in the same manner as he cleans his slides, by wiping them with a weakly acid alcohol solution. Each slide is taken individually, drained by its corner, and laid on a flat surface; a drop of mounting medium is placed on its top. Then the coverslip is placed on the mounting medium and pressed down with a needle. It should not be pressed absolutely into contact with the slide, or too thin a layer of mounting medium will be left. Some experience is required to judge when the coverslip has been pushed down far enough. If this is done skillfully, the surplus mounting medium will form a neat ring around the outer surface of the cover. If it does not do so, care should be taken that no portion of the cover is devoid of surplus mountant that will be sucked under the coverslip as the solvent evaporates. These slides should be left to dry at room temperature for about 1 day and then placed on a warm plate for about 1 week. After they are dried, the surplus dry mounting medium should be scraped off with a knife and the excess, remaining after scraping, removed carefully with a rag moistened in 96 per cent alcohol.

SUMMARY

1. Bleed an anesthetized frog to death. Fix short lengths of intestine in Helly's fluid overnight.
2. Wash specimen in running water for some hours. Dehydrate, clear, and embed in wax.
3. Trim block and attach to microtome. Cut ribbon of sections and mount single sections or short lengths on clean slides, using diluted Mayer's albumen to flatten sections. Dry thoroughly.
4. Heat dried slides until wax melts. Place in xylene until all wax is removed. Remove xylene with absolute alcohol and transfer slides to water.
5. Transfer to celestine blue B solution for 1 min. Wash in water.
6. Transfer to 1 per cent eosin until sections appear stained sufficiently. Wash in water.
7. Dehydrate, clear, and mount in balsam.

EXAMPLE 13 | Preparation of a Transverse Section of a Stem of Aristolochia Stained by the Method of Johansen

In the preparation of a transverse section of a root described in Example 11, nothing was required except that an outline of the cells be shown as a skeleton and that this skeleton be differentiated as to woody and nonwoody tissues.

The present example describes the preparation of a section of a plant stem preserved, sectioned, and stained in such a manner as to differentiate all of the components present. The technique to be described could be applied equally well to a leaf or any other part of a plant; the stem of an Aristolochia has been selected only because of its wide use in teaching.

First it is necessary to collect and fix the material. The choice of fixative is not of any great importance, but the majority of workers prefer a Craf fixative, such as the formula of Lavdowsky given in Chapter 6, for this type of material. About a pint of this solution should be prepared immediately before it is required for use.

Great care must be taken in the selection of a stem. If a second- or third-year growth is taken, it will be too woody to cut satisfactorily, while a new growth of early spring will not be differentiated sufficiently to justify applying a complex stain of the type used in this example. The best Aristolochia material is usually obtained from the new season's growth when it has partially ripened toward the end of August. Care must be taken that the piece of stem is not so cut as to permit the introduction of air into the transporting vessels since this will greatly increase the difficulties of subsequent manipulation. Unless the stem can be cut in the early morning while it is completely turgid, it is best to pull down a branch of the vine and make the first cut under the surface of water in a bucket. About a foot of stem is pushed under the water and cut off. The piece should be transferred immediately to a large volume of fixative

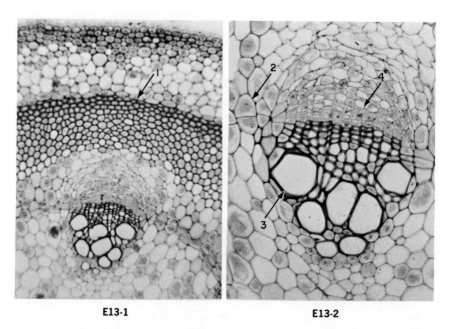

E13-1 E13-2

Figs. E13-1 and E13-2 | Details from an excellent preparation produced by the method described in this example. Fig. E13-1 shows a general view of the stem. Notice at 1 that the lignified band is sharply defined and has remained attached to the outer cortical cells. Fig. E13-2. Details of a vascular bundle. Notice at 2 the relatively unshrunk cytoplasm showing a clear nucleus. The sharply defined wall of the vessel is shown in 3, and 4 shows unshrunk well-stained phloem.

and stored in the dark for about 3 days. At the end of this time the central portion of the stem should be cut into ½-in. lengths with care so as to avoid crushing, and the pieces washed overnight in running water.

The process of dehydrating and clearing plant material is very different from that used for animal tissues. Unless the greatest distortion of the cell contents does not matter, it is necessary to employ the long series of dehydrating mixtures given in Chapter 8, and it is recommended, at least for the beginner, that he use two changes of tertiary butyl alcohol as the last step. The tube containing the object in tertiary butyl alcohol now has shredded into it enough of the embedding medium to be used to fill about one-tenth of the tube. It is left in this condition overnight and then about as much more wax added. This process is continued until there is rather more wax in the tube than there was originally tertiary butyl alcohol. It must be emphasized that this method of slow impregnation is just as important for plant tissues as slow dehydration and clearing. The tube containing the sludgy mixture of wax and tertiary butyl alcohol is now placed in the oven and left for 4 or 5 hr. The specimens are then carefully removed to a container of pure molten wax. One hour in this bath would be enough for a small section of stem. The piece is then transferred to a second bath of pure paraffin for a further hour before being made into a block. The method of preparing the block and cutting sections from it is identical with that used for animal tissues, and the description given in Example 12 may be followed without alteration.

It will be noticed by reference to Chapter 7 that Johansen's stain requires both methyl Cellosolve and tertiary butyl alcohol. Neither of these is a common laboratory reagent, and their availability should be checked before commencing the technique. There will be required four staining solutions (safranin, methyl violet 2B, fast green FCF, and orange G), three differentiating solutions (which consist of mixtures of various solvents in proportions that must be followed closely), and one special dehy-

Figs. E13-3 and E13-4 | Details from a bad section prepared by the method given in this example. The section appears to have been cut through a node, so that the arrangement of the parts is hopelessly distorted. The separation of the layers seen at 1 in Fig. E13-3 (compare with 1 in Fig. E13-1) is a result of both improper dehydration and a blunt knife. The folded tissue seen at 2 is caused by imperfect flattening. The blur at 3 (Fig. E13-4) is caused by an air bubble left under the section. Notice the imperfectly stained wall of the vessel at 4 (compare with 3, Fig. E13-2); at 5 there is hopelessly shrunken cytoplasm (compare with 2, Fig. E13-2).

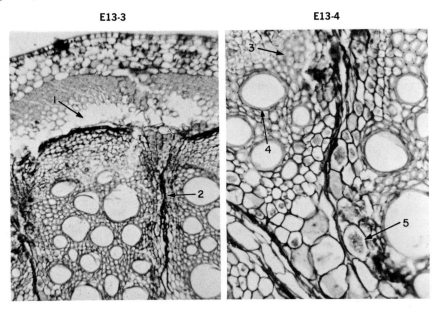

drating solution that must be used to avoid removing the stain in the final dehydration.

The worker thus should have in front of him the following coplin jars of reagents arranged in two rows: The jars in the front row should contain xylene, absolute alcohol, 95 per cent alcohol, and 70 per cent alcohol. Those in the back row should contain, in order, the safranin solution, the methyl violet solution, the first differentiating solution, the fast green FCF, the second differentiating solution, the orange G, the third differentiating solution, and the special dehydrating reagent. A large dish should be available in the sink, in which the slides can be washed in running water.

The description to follow will be based on the handling of a single slide, although there is nothing to prevent racks of slides from being taken through as readily as a single example.

A slide is heated over a flame until the wax is melted. Then it is placed in xylene until the whole of the wax has been dissolved. It is transferred to absolute alcohol for the removal of the xylene, to 95 per cent alcohol for the removal of the xylene-contaminated absolute alcohol, and then to 70 per cent alcohol. The slide is placed in safranin and left from 1 to 3 days or until the nuclei are seen to be deeply stained in red. Then the slide is washed in running water until no more color comes away. This initial staining in safranin is the point at which the technique is usually interrupted since the slides may be left in the staining solution indefinitely without damage, whereas the subsequent operations must be conducted consecutively.

The thoroughly washed slides are transferred to the methyl violet solution for 10 to 15 min. On removal, they should be rinsed briefly in running water to remove the surface stain before being differentiated in the first differentiating solution from 10 to 15 sec. Then they are placed directly into the green staining solution, where they remain from 10 to 20 min or until their red-purple color is changed to a greenish hue. The slide is taken from the green solution and drained from the corner, and the back of it is wiped. A very brief rinse (5 to 10 sec) in the second differentiating solution precedes the transfer to the clove oil–orange G solution. It is perfectly satisfactory to have this solution available in a drop bottle from which it can be poured on the slide, although if any large number of sections are to be handled, a coplin jar is usually more convenient.

The slide should be removed from the orange stain at intervals and examined under a microscope. Staining should be discontinued when the cytoplasm of the cell changes from a muddy gray color to a clear bright orange. This will normally take from 3 to 5 min, although occasionally a longer period of time is required. As soon as the required color has been reached, the slide is dipped up and down three or four times in the special dehydrating agent, the primary purpose of which is to remove the excess orange from the surface of the slide. Then the slide is transferred to xylene, in which the special dehydrating agent should be removed completely, before being mounted in balsam in the ordinary manner.

SUMMARY

1. Select a suitable Aristolochia stem and fix in Lavdowsky's fluid.
2. Wash short lengths of stem in running water.
3. Dehydrate through the mixtures of water, alcohol, and tertiary butyl alcohol given in Chapter 8.
4. Accumulate the pieces in pure tertiary butyl alcohol and add wax little by little.
5. Transfer to oven and complete embedding process through two changes of pure wax.
6. Make block and cut sections in the manner described in Example 10.
7. Mount individual sections or short lengths of ribbon on slides, dry, dewax, and transfer to 70 per cent alcohol through the usual reagents.
8. Stain in Johansen's safranin from 1 to 3 days.
9. Wash thoroughly in running water and stain in 1 per cent methyl violet 2B for 10 to 15 min.
10. Rinse in water and differentiate 10 to 15 sec in the first differentiating solution.
11. Stain in fast green FCF from 10 to 20 min.
12. Rinse in second differentiating solution 5 to 10 sec and stain in orange G 3 to 5 min.
13. Rinse off orange G in third differentiating mixture, remove this mixture with special dehydrating agent, rinse in xylene, and mount in balsam.

EXAMPLE 14 | Demonstration of Spermatogenesis in the Rat Testis, Using the Iron Hematoxylin Stain of Heidenhain

The laboratory white rat is one of the best forms in which to show spermatogenesis because it has a continuous breeding period, so all stages are available in almost every section examined.

The rat selected should be a young male. It can be killed most conveniently with chloroform. The scrotal sac is opened by a median incision, and the testes are removed, trimming away the epididymis. The testes should be placed on a clean glass plate and slashed with a sharp scalpel or razor about two-thirds of the way through by cuts spaced a few millimeters apart before being thrown into the fixative solution selected. Few fixatives for this purpose can surpass Allen's fluid (Chapter 6). At least 100 ml of fixative should be employed for a normal-sized testis, and the bottle containing it should be reversed once or twice during the first few hours to avoid the accumulation of diluted fluid at the bottom. The time of fixation is not of any great importance but, in any case, should be overnight and, in general, should not exceed a few weeks. After the object is removed from the fixative, it may be washed for about an hour in running water before being transferred to 70 per cent alcohol to complete the removal of the picric acid. It must be emphasized that washing a specimen in water, after picric acid fixatives, results in a great deal of vacuolation of the cytoplasm, although this does not, in the present instance, interfere with the object being studied. After three or four changes of 70 per cent alcohol—the testes remain in a large volume of solution for at least 2 or 3 days between changes—the final removal of the picric acid may be accomplished by adding a small quantity of dry lithium carbonate to the alcohol used for washing. It will be impossible to remove all the yellow color, some of which is occasioned by the combination of the picric acid with the albuminoids present, but the last alcohol used for washing should be only very faintly tinted with yellow. It is not the color of the fixed material to which objection is raised during the passage of the material through paraffin; it is the fact that, unless most of the picric acid is removed, there will be crystallization with consequent damage to the tissues.

Small pieces may now be removed for embedding from the testis itself. It is best to select pieces about 1 mm in from the surface and from a side of about 2 mm. These should be embedded in paraffin and cut about 5 μ thick in the usual manner. Then the sections should be attached to slides, particular attention being paid to the fact that the slides are clean and that not too much Mayer's albumen, which may interfere with differentiation or staining, is used.

The dry sections attached to the slides should be warmed on the underside until the paraffin melts, placed in xylene until the paraffin is removed completely, and then run down in the ordinary way through absolute alcohol and lower-percentage alcohols to distilled water. Then they should be lifted from the distilled water and examined carefully. If there is a tendency for the water to gather in droplets on the slide or, if upon shaking the water from the slide, each section appears to retain an adherent coat of water around itself, it is an indication that either the wax was not removed properly in the xylene or the xylene itself is so old as to have a wax content too high to be useful. Such slides must be returned through the alcohols to absolute alcohol and thence to clean xylene, where they should be left for a few minutes before again being brought down to the water and reexamined. There is no more common cause of the failure of the stains to take than the imperfect removal of the wax.

Only two solutions are required for staining. These are a 2.5 per cent solution of ferric alum and a 0.5 per cent solution of hematoxylin. The only difficulty in making the ferric alum solution is to obtain a pure and unoxidized sample of the reagent. Most of the crystals in a new bottle are of a clear violet color, but after a bottle has been opened for some time, particularly if the stopper is loose, most of the crystals become covered with a brownish deposit that must be scraped off with a knife before the solution is prepared for staining. If the brown powder on the outside of the crystal forms a layer of any thickness, it is best to reject the whole and get a fresh supply of the reagent. Hematoxylin itself has little staining effect; the color is produced by the formation of lakes with hematein, an oxidation product of the hematoxylin. It was customary in former times to prepare large quantities of solution, which were kept with the stopper loose in the bottle for a period of at least 1 month before use. For the purpose of Heidenhain's technique, however, it is far more important that a small quantity of the ferric alum be carried over into the hematoxylin solution than that the latter be aged. The staining will be found both simpler and more effective if a few drops of hematoxylin are placed in the iron alum solution and a few drops of the iron alum solution are placed in the hematoxylin. Both solutions, of course, should be filtered immediately before use if the finest slides are required because chromosome figures in a rat can be obscured by a very small particle of dust.

The slides are next taken from distilled water and placed in the mordant solution. It matters little how long they remain in this solution, although the usual directions call for keeping them there overnight. This varies with every type of tissue studied and is greatly dependent on temperature. If the solutions are heated to $50°C$, with the understanding that this will cause a swelling of the section and a general obscuring of the finer details, the period may be shortened to as little as 10 min. The finest stains, however, are those obtained by leaving the sections in the mordant solution at room temperature overnight. On removal from the mordant solution, the sections should be rinsed *very* briefly in distilled water; the purpose of the rinse is to remove the surplus mordant from the surface of the slide without extracting it from the tissues. The slides are then placed in the hematoxylin solution, where they should remain for approximately the same period as they have been in the mordant; the length of time is not important, although from 3 to 24 hr is the customary period. Sections may be removed from time to time from the staining solution and examined with the naked eye. Successful preparations show the sections to have been blackened completely, although a slight bluish tinge in the black is permissible. If they have not become blackened completely in 24 hr, it is necessary only to place them, after a brief rinse, in the mordant solution and leave them there for another 24 hr before returning them to the stain.

If, however, the sections are sufficiently blackened on removal from the staining solution, they may be differentiated, that is, the hematoxylin stain extracted from all portions of the sections except the chromosomes, which are to be studied. This is done customarily with the same solution in which they were mordanted, although, of course, a fresh solution or a stronger solution may be used if desired. Differentiation at the commencement of the process goes relatively slowly, so that all the slides, which are carried in a glass rack, may be removed and placed in the 2.5 per cent iron alum without any very great care. The actual time in which differentiation takes place cannot be forecast since it depends on a large number of uncontrollable factors; it is never less than 5 min, nor is it very often more than a few hours. Sections, therefore, should be withdrawn from the iron alum every 4 or 5 min and examined briefly under the low power of a microscope. It is a matter of great convenience for the controlling of differentiation of chromosomes in this type of preparation if an ordinary student microscope (of the type customarily used in the laboratory) can be lifted with a glass plate over the stage, so that one can, without fear of damage to the instrument, place a slide, wet with iron alum, on the surface of the stage for examination. No more common accident takes place than the placing of the slide upside down on the surface of the stage with the subsequent loss of all the sections. This will be avoided readily if the worker will make it a matter of routine, as he lifts the slide from the mordant, to hold it at an angle between himself and a light source so that the light is reflected from the surface. If the sections are, as they should be, on the upper surface of the slide when it is so placed, they will appear to be double through the reflection from the under surface as well as the upper surface of the glass. A good rule is never to place a slide which has just been taken from a fluid on the stage of the microscope until one has seen the double reflection.

If a low-power examination of the section shows the nuclei to be standing out clearly, the entire tray should be removed to distilled water because from this time on differentiation is very rapid, and each slide must be controlled separately. If, however, the nuclei are not sharply defined and a considerable degree of black or bluish color remains in the background, then the entire tray may be left in the iron alum for as long as is necessary. When this preliminary differentiation down to the distinction of the nuclei under low-power examination has been completed, it is necessary to continue differentiation while examining the slides at frequent intervals under a very high power of the microscope. It is convenient to have available a water-immersion objective. It is obviously impossible to place immersion oil on a wet slide, and the short working distance of a high-dry objective renders it particularly liable to cloud up from the evaporation and recondensation of the water. Water-immersion objectives are usually of 3-mm equivalent focus, and this gives a sufficiently wide field to permit differentiation to be observed, while at the same time it has sufficiently high magnification for satisfactory control. Each slide should be taken separately, returned to the 2.5 per cent iron alum for a few minutes, and then reexamined. The various phases of mitosis and meiosis do not retain the stain to the same degree, and care must be taken that the color is not washed completely out of the other chromosomes as a result of examination only of metaphase figures, in which the color is retained longer than in any other. A great deal of practice is required to gauge accurately the exact moment at which to cease differentiation, which may be stopped almost instantly by placing the slides in a slightly alkaline solution. In Europe most tap waters are sufficiently alkaline for the purpose and are generally specified, but in the cities of the United States it is often best to add a very small quantity of lithium carbonate or sodium bicarbonate to the water used to stop differentiation. Slides may be left in this for any reasonable period of time; the process is complete when the slide turns from a brown to a blue color.

Then the slides are rinsed in distilled water, graded up through the various percentages of alcohols, dehydrated, cleared, and mounted in balsam in the usual manner. A section on a slide, which is required for examination over a long period of time, should be some distance from the edge of the coverslip, for, as the balsam oxidizes inward from the edge, it tends to remove the color of the stain from the chromosomes, leaving them a rather unpleasant shade of brown. If this happens to a valuable slide, however, the matter can be remedied by the utilization of a green light, which will make the chromosomes appear black again.

SUMMARY

1. Remove the testes from the scrotal sac of a young male rat and trim away the epididymis.
2. Slash the testes deeply with a razor, making cuts 2 or 3 mm apart. Fix in at least 100 ml of Allen's fluid from 1 to 7 days.
3. Wash for about 1 hr in running water before removing small pieces to 70 per cent alcohol. Change alcohol daily until no further color comes away.
4. Dehydrate, clear, embed in wax, and cut $5\text{-}\mu$ sections. Attach sections to slide. Dewax dried slides and grade down to water through usual reagents.
5. Mordant overnight in 2.5 per cent iron alum.
6. Rinse briefly and stain overnight in 0.5 per cent hematoxylin.
7. Rinse briefly and place in 2.5 per cent iron alum until chromosomes are clearly differentiated.
8. Wash in alkaline tap water until sections are blue. Dehydrate, clear, and mount in the usual manner.

EXAMPLE 15 | Preparation of a Transverse Section of the Tongue of a Rat, Using Celestine Blue B Followed by Picro Acid Fuchsin

The chief difficulty in preparing a transverse section of the tongue is to avoid the hardening of the muscle, which tends to become brittle if either imperfectly fixed or handled with undesirable reagents in any stage of the proceedings. Therefore, it is recommended that the following description be followed rather closely, for it can be adapted almost without variation to any other heavily muscularized tissues which it is desired to stain.

The tongue may be removed most easily by severing the articulation of the lower jaw and removing this together with the adherent tongue, which may be detached with a short scalpel or cartilage knife. A portion of the tongue, approximately 5 mm in length, is cut off and placed in a large volume of the selected fixative.

Although opinions vary widely as to the most desirable fixative to employ for muscularized tissues, it may be said at once that no alcoholic solution or solution containing picric acid or mercuric chloride can be employed under any circumstances. The author's choice would be the solution of Petrunkewitsch, which he has employed most successfully on a variety of heavily muscularized tissues. This formula would also provide an excellent premordanting for the staining techniques that follow. Whatever formula is selected, however, a very large volume should be employed and permitted to act for no longer than is necessary to secure the complete impregnation of the tissues. If the operator does not wish to employ a copper formula, it may be suggested that he use any formula containing nitric acid or one of the weaker dichromate mixtures. When the piece has been fixed successfully, it must be washed overnight in running water and then dehydrated. It is during the process of dehydrating, clearing, and embedding that most muscularized tissues become unmanageable. Nothing, of course, can counteract the effect of improper fixation, but even with good fixation the utmost attention must be paid to the selection of dehydrating agent and clearing agent and to the temperature at which the embedding takes place. It has been the experience of the author that the newer substitutes for alcohol in dehydrating tend to harden or render brittle muscular tissue to a greater extent than the more old-fashioned method of using ethyl alcohol. There is little choice in the matter of clearing prior to embedding, for it has been found by numerous workers that benzene has less hardening effect on muscular tissue than any other agent.

Unless very thin sections are to be cut, it is recommended that a wax of no higher melting point than 52°C be employed and that embedding be conducted exactly at this temperature. It is far easier to do this by the method of an overhead heating light placed above a tube of paraffin than by thermostatic control. It may be stated categorically that should the temperature be permitted to rise above 56°C, it would be better to throw the preparation away than to waste time endeavoring to section it. Paraffin sections are cut from the block by the standard method, stranded, and caused to adhere to the slide by egg albumen. These sections, however, may be lost, for muscularized tissues tend to absorb water very readily when in paraffin section, so that, if the sections are flattened for a prolonged period in contact with large volumes of warm water, they will tend to expand more than the surrounding wax with the result that they will arch away from the glass support and inevitably become detached while being deparaffinized. Therefore, it is recommended that, as soon as the sections are flattened, they be pressed to the slide with a piece of wet filter paper, rolled into position with a rubber roller, and dried with the maximum possible speed.

As soon as they are dried, the sections are deparaffinized by the usual techniques and taken down to distilled water, where they can remain until one is ready to stain them. Celestine blue B is selected as the nuclear stain in this instance because the contrast of muscularized tissues is far better brought out with the aid of a picrocontrast than by any other method. These picrocontrasts are, however, so acid that hematoxylin-stained nuclei are often decolorized in the course of counterstaining. The sections are therefore passed directly into celestine blue B staining solutions (Chapter 7) from the distilled water and allowed to remain 1 min. After they are removed from the staining solution, they are rinsed in distilled water and accumulated in a jar of either distilled or tap water until it is time to counterstain them.

The solution of van Gieson should be used for counterstaining. This stain requires little or no differentiation, so that the sections may be placed in it and examined from time to time until the connective tissues are seen to be stained a very bright red against a background of yellow muscle. If a small quantity of red is picked up by the muscle fibers, it will be removed in the process of differentiation. The time is not critical, but that given in the formula cited (2 to 10 min) will be found to cover the range normally necessary.

A slight difficulty will be occasioned in dehydration through the tendency of the picric acid to leave the tissues in the various alcohols employed. Either this may be prevented by dehydrating in a series of 1 per cent solutions of picric acid in the various alcohols or it may be ignored completely according to the depth of yellow color that is to be retained. If, on the contrary, a very pale yellow is desired, the sections may have to be put in 96 per cent alcohol, for a period of time beyond that necessary for dehydration, to remove the unwanted picric acid. Then the sections are cleared in xylene in the normal manner and mounted in balsam.

SUMMARY

1. Place a 5-mm slice of tongue in 100 ml of Petrunkewitsch's fluid for 1 to 2 days.
2. Wash overnight in running water. Dehydrate, clear, embed in low-melting-point wax, and cut 10- to 12-μ sections.
3. Flatten sections and squeeze to slide. Dry as rapidly as possible. Dewax dried sections and grade down to water through usual reagents.
4. Stain in celestine blue B for 1 min. Overstaining is not possible.
5. Rinse in water and stain in van Gieson's picro acid fuchsin from 2 to 10 min.
6. Dehydrate and differentiate in 95 per cent alcohol. Clear and mount in the usual way.

EXAMPLE 16 | Preparation of a Transverse Section of Amphioxus, Using the Acid Fuchsin—Aniline Blue—Orange G Stain of Mallory

Amphioxus is a difficult subject from which to prepare satisfactory sections, the more so as it is almost impossible nowadays to secure supplies of living Amphioxus and to fix them oneself or to prevent the supplier from whom one obtains the fixed material from using Bouin's fluid. If it is possible to get the living animals, they should be fixed by the methods recommended for heavily muscularized material in Example 15. If this is not done, it is almost impossible to obtain unbroken sheets of muscle in the transverse section unless one is prepared to sacrifice a certain amount of histological detail in the interest of morphological demonstration. It is also greatly to be regretted that popular demand has forced the biological supply houses to sell only the very large specimens, with the result that the sections are too large to be viewed at one time in even the lowest power commonly available on a student microscope. If any selection can be exercised, care should be taken to pick a specimen of not more than 2.5 mm thickness, so that it may be seen as a whole.

If, however, one is forced to use a Bouin-fixed specimen, it may be sectioned without too much difficulty, provided that it is first soaked overnight in 1 per cent nitric acid. This treatment naturally destroys much of the fine cytological detail and should not be applied to any specimen in which it is desired to demonstrate, for example, the detailed structure of the endostyle. The author, however, is always prepared to sacrifice such detail as this in a section desired for class demonstration to avoid having to answer endless questions as to what is this and that cavity, which will be seen in a section of Bouin-fixed Amphioxus handled by routine methods.

Apart from this question of fracturing of the muscular layers, no difficulty will present itself in sectioning, and as many 10-μ sections as are required should be accumulated. If it is desired to place on the same slide a collection of sections from different regions of the animal, reference may be made to the description of this procedure in Example 18.

When the sections have been mounted on slides, deparaffinized, and graded down to water, it is recommended that they be treated overnight in a saturated solution of mercuric chloride and then washed in running water for at least 6 hr. This process improves the vividness of Mallory's stain almost beyond belief when it is applied to a section of Bouin-fixed material. The solutions used for Mallory's stain present no difficulty in their preparation, although it is recommended that 1 per cent phosphotungstic acid be substituted for the 1 per cent phosphomolybdic acid specified in the original method. The sections, when they have been washed thoroughly after the mordanting in mercuric chloride, are placed in the 1 per cent solution of acid fuchsin for a period of about 2 min. This time is not critical; it is necessary only to make sure that the entire section is thoroughly stained. On removal, the sections are rinsed rapidly in water for the purpose of removing the surplus stain and then are placed in the 1 per cent phosphotungstic acid until such time as the red stain has been removed entirely from the connective tissues. This may be judged partly by the cessation of the color clouds that rise from the section and partly by an examination under the lower power of the microscope to make sure that the septa between the myotomes are free from color. The specified time of 2 min is usually sufficient, and the sections will not be damaged however long they may be left. On removal from this solution, they are rinsed quickly in water and placed in the acid–methyl blue–orange G solution, where they should remain for at least 15 min. The mistake is often made of leaving them for too short a time in this stain, for they will have the appearance of being deeply stained after an immersion of only a few moments. It does not matter how long they remain; it is the author's experience that soaking for at least 15 min discourages the subsequent removal of the blue from the tissues. After they are removed from this rather thick staining mixture, the slides are washed thoroughly in water. This wash is designed not only to remove the whole of the adherent stain from the slide but also to permit the oxalic acid to be leached out of the tissues. No differentiation of the stain should take place in water, since the necessary differentiation is produced by the absolute alcohol used for dehydration in the next stage. It is quite impossible to take sections stained by this method up through the common graded series of alcohols, but no grave damage will be occasioned by the omission of this step. If, however, the operator is one who insists that his sections pass through a graded series, mixtures of acetone and water should be substituted for alcohol and water. When the sections reach absolute alcohol, they should be watched very closely, while being moved continuously up and down in the alcohol. The blue color will leave them in great clouds; these clouds will taper off rapidly, leaving a terminal point at which no color leaves for a moment or two before a slow stream again starts to appear in the alcohol. As soon as the initial color clouds cease, the sections should be removed to xylene, which instantly stops the differentiation. If the operator is uncertain of this method or is trying it for the first time, it is recommended that the slides be washed thoroughly in absolute alcohol but removed to xylene before the color clouds have ceased to leave the sections. Examination under the low power of the microscope will show these preparations to have dull purple muscle tissue and intensely blue connective tissue. A few trial sections should now be returned to absolute alcohol for a few moments, put back into xylene, and then reexamined. It is possible by this means to exercise the most perfect control over the differentiation, which should be stopped when the muscles and nuclei are clear red and the connective tissues are a clear light blue. No attention should be paid, while differentiating, to any of the structures, such as the gonad, which by this method acquire a violet coloration that cannot be judged; the process should be controlled only by apparent contrast between the pure blues and pure reds in the section.

The stains used in this preparation are alkali-sensitive, and it is a customary procedure in Europe to mount them in as acid a medium as possible. If one is using one of the synthetic resins, which are neutral, it is strongly recommended that the coverslip, before it is applied to the resin, be dipped briefly in a strong solution of salicylic acid in xylene. This salicylic acid will dissolve in the resin and provide a permanently acid medium; with this precaution, the sections are permanent.

SUMMARY

1. Fix a small adult Amphioxus in any good fixative. If only a Bouin-fixed specimen is available, soak it overnight in 1 per cent nitric acid.
2. Cut 10-μ paraffin sections in the usual way, attach to slide, and grade to water.
3. If fixative does not contain mercuric chloride, soak sections overnight in a saturated solution of this salt and wash for some hours in running water.
4. Stain in 1 per cent acid fuchsin for about 2 min.
5. Rinse quickly in distilled water and transfer to 1 per cent phosphomolybdic acid until no more color comes away (about 2 min).
6. Rinse quickly and place in Mallory's oxalic acid–methyl blue–orange G mixture for 15 min.
7. Wash thoroughly in water. Rinse in 95 per cent alcohol to remove adherent water. Differentiate each slide individually in absolute alcohol. Stop differentiation with xylene and mount in balsam.

EXAMPLE 17 | Demonstration of Diplococci in the Liver of the Rabbit, Using the Phloxine—Methylene Blue—Azur II Stain of Mallory, 1938

This method of Mallory is the best of all the eosin–methylene blue methods that have from time to time been suggested for staining bacteria in sections. It has the advantage of giving a first-class histological stain, in addition to differentiating bacteria, and it might well be used as a standard procedure in place of the more customary hematoxylin-eosin, at least in pathological investigations. The liver of a rabbit is so frequently infected with diplococci that it has been selected as a type demonstration, for such infected animals will be found in ordinary laboratory investigations, making it unnecessary to go to the trouble of infecting a rabbit for the purpose of obtaining the necessary demonstration material.

If, then, in the course of routine dissections, a sacrificed rabbit is found to have a pneumococcal infection of the liver, which may easily be seen as yellow lesions on the surface, it is necessary only to cut the lesion and some surrounding tissue and to place it in a suitable fixative. Mallory himself recommends fixation in Zenker's fluid for about 24 hr. As Zenker's fluid contains mercuric chloride, it is undesirable that the specimen remain in it for more than 3 or 4 days, but the actual time of fixation is not critical. As always, when dealing with dichromate fixatives, a large quantity of fixative should be used, and the object should be suspended in the fixative solution in a loose cloth bag. When fixation is complete, the pieces are removed from the fixative and washed in running water overnight. They are then embedded in paraffin and sectioned in the ordinary manner. Sections of from 5 to 8 μ are desirable if it is intended to demonstrate the bacteria, although these are readily seen in the 10-μ sections customarily employed for histological examinations. When the sections have been fixed on the slide, they may, if desired, be freed from the last traces of mercuric chloride by treating them with iodine and bleaching with sodium thiosulfate. The writer does not usually do this, but the treatment is insisted upon by Mallory in the description cited.

The staining solutions used in this technique are simple to prepare and relatively stable. The first solution is a 2.5 per cent solution of phloxine in water. Two stock solutions are also required: one of 1 per cent each of methylene blue and borax in distilled water, and the other a 1 per cent solution of azur II in distilled water. Five milliliters of each are added to 90 ml of distilled water for use. Differentiation is in Wolbach's resin alcohol.

It is difficult to get a sufficiently heavy stain in phloxine to withstand the alkaline thiazine solutions used for counterstaining. Mallory recommends that the sections be placed in a coplin jar of the solution in a 55°C oven and that they remain there for at least 1 hr. The coplin jar is then removed from the oven and cooled before the sections are removed; they should be stained a dense orange. If they have not yet acquired this color, they should be returned to a paraffin oven for a further period. If the sections are satisfactorily stained, the solution may be poured off or the slides removed from it and briefly rinsed in water. The purpose of this rinse is not to differentiate the eosin in the section but to remove it from the glass. The slides bearing the sections are then placed in the methylene blue–azur solution in another coplin jar for 5 to 20 min; the exact time varies according to the specimen that one is staining and can be determined only by experiment. Mallory recommends that the solution be freshly filtered onto each side, rather than that the solution be used in a coplin jar, but the writer has not found this nearly so convenient, nor does it appear to be in any way obligatory. After the slides have taken up sufficient methylene blue solution to appear bluish rather than pinkish, they may be accumulated in water, before being differentiated in the resin alcohol one at a time. This differentiation is best conducted in a dish that is large enough to admit the slide in a flat position. The slide is taken in a pair of angle forceps, dipped under the surface of the solution, and waved gently backward and forward for about 1 min. As it is being moved backward and forward in the differentiating solution, the blue color will come off in clouds; it is much easier to overdifferentiate than to underdifferentiate.

The differentiation can be readily controlled by inspection under the microscope, although it is not necessary to observe the bacteria, since the nuclei have exactly the same staining reaction. Differentiation should be stopped when the nuclei can be seen, under the low power of the microscope, to be very deep blue, while the general background of the section is pink. After a little practice the required color may be gauged without examination under the microscope.

These specimens are not permanent unless the alcohol and resin are removed from them. It is desirable, therefore, as soon as differentiation is complete, to dehydrate them in absolute alcohol and then clear them in at least three changes of xylene to make sure that no alcohol can be carried over into the mounting medium.

SUMMARY

1. Fix diplococcus-infected liver in Zenker's fixative.
2. Cut 5- to 8-μ sections. Mount on slide and run down to water through usual reagents.
3. Stain 1 hr in 2.5 per cent phloxine at 55°C.
4. Rinse briefly and transfer to Mallory's methylene blue–azur for 5 to 20 min.
5. Wash in water. Differentiate in resin alcohol.
6. Wash thoroughly in absolute alcohol.
7. Clear in two or three changes of xylene and mount in balsam.

EXAMPLE 18 | Preparation of a Series of Demonstration Slides, Each Having Six Typical Transverse Sections of a 72-hour Chick Embryo, Using the Acid Alum Hematoxylin Stain of Ehrlich

Example 3 described in some detail the manner in which a chick embryo can be removed from the yolk and fixed in a syracuse watch glass, where it is stretched by a collar of filter paper during fixation. Exactly the same procedure should be followed in the present instance, save that it is not necessary to leave the hole in the paper of a size larger than will accommodate the embryo itself. The same fixative recommended in Example 3 should be employed, and, after the removal of the fixative, the embryo should be embedded in paraffin by the technique described in Chapter 12. Then the complete series of serial sections should be taken throughout the whole embryo, and the ribbons should be accumulated on a sheet of black paper in front of the worker.

It is presumed for the purpose of this example that the reader desires to prepare a series of slides for class use on each of which there will be arranged, in order, transverse sections through the regions of the eye, ear, and heart and through the anterior, middle, and posterior abdominal regions. In these regions will be found all that is required for the purpose of teaching an elementary class the development of the eye, ear, and heart and the closure of the amnion and neural folds. It is necessary first to identify those sections that will show the required structure and isolate the portions of ribbon containing them. Provided that the sections are placed against a background of black paper, this is relatively simple with the aid of a long-arm binocular dissecting microscope, which may be swung over the ribbons and which will supply sufficient magnification to enable the regions of the ribbon to be identified by a competent microtomist. If the operator has had little practice at this, it might be desirable to stain the embryo in Mayer's carmine before embedding. Then each portion that contains the selected sections is cut from the ribbon with a sharp scalpel (which is moved with a rocking motion), picked up on a camel's-hair brush, and transferred to another sheet of black paper. The remains of the ribbon may be thrown away.

The sections in each of the selected strips of ribbon are counted to determine the maximum number of slides that may be made—the ear sections are usually the limiting factor—and the pieces of ribbon trimmed, each to contain approximately the same number of sections. Then the required number of slides are cleaned and a few drops of the usual adhesive added to 25 ml of filtered distilled water in a small flask.

The only difficulty of the procedure consists in persuading each section to occupy its correct place on the slide. This is carried out most easily by the following technique: A single slide is taken, placed in front of the operator, and covered lightly with the diluted adhesive. The fluid should extend to the edge of the slide but should not be raised in a meniscus sufficiently high to cause any appreciable slope of the fluid from the center of the slide toward the edges. Using a sharp scalpel, the end section is cut, with a rocking motion, from each of the ribbons. These sections are placed in the correct order but without any regard to symmetry on the surface of the fluid on the slide. To secure these sections in the required position, it is necessary to have two fine brushes, a mounted needle, and a bunsen burner or spirit lamp.

The last section (that is, the section which is to lie farthest from the label on the slide) is now secured in position with a brush held in the left hand, while the second section is maneuvered with a brush held in the right hand until its edges touch those of the first section. Both sections will be held together by capillary attraction when the brush is removed. The needle is warmed in the flame and used to fuse the edges of the sections together in two spots. If the entire edge is melted, it will cause a ridge which will prevent the compound ribbon from lying flat against the slide; two minute spots fused together with the point of the needle are sufficient to hold the section in place. The needle is laid down and the brush again picked up with the right hand and used to guide the next section into its appropriate place. This section is spotted into position with the tip of the warm needle, and so on, until all the sections have been fused into a continuous ribbon. Then the slide is placed in the usual way upon the warm table until such time as the ribbons have flattened; they are drained in the manner described in Chapter 12, and then pressed to the slide with a wet filter paper and rubber roller, if this is the method of operation preferred by the technician. The compound ribbon, of course, may be guided into the center of the slide, while the latter is still wet, before it is pressed or dried. The sections are left, in the ordinary course of events, on the warm table until they are entirely dry before being dewaxed in xylene and brought down to 90 per cent alcohol through absolute alcohol in the usual manner.

Ehrlich's acid alum hematoxylin has been selected for this typical example because it is one of the best, although at the same time one of the most frequently misused, of the hematoxylin stains. The method given for its preparation should be rigorously followed; that is, the hematoxylin should be dissolved in a mixture of acetic acid and absolute alcohol, and then the glycerin, water, and ammonium alum should be added to the bottle, which should be shaken vigorously and allowed to ripen with the stopper loose for some months. Artificially ripened hematoxylin does not give so good a preparation, but there is no reason why this stain should not be prepared in half-gallon lots at routine intervals, so that a sufficiently ripened solution is always available. When it has once been ripened, which can be told both by its fruity smell and dark color, it remains in a fit condition to use for many years. One of the most frequently omitted precautions is the maintenance of the concentration of the ammonium alum by the addition of about 100 g per liter to the bottle after it has been sufficiently ripened. This stain should never be diluted but should always be used full strength by the method now to be given.

All the slides, gathered together on a glass tray (each slide may be treated individually), are taken from the 96 per cent alcohol and placed in full-strength Ehrlich's hematoxylin solution for a few minutes. The exact time is not important, but they should be examined at intervals to make sure that they are not becoming overstained. In the absence of experience, it is recommended that a period of 1 min be used and that then they be examined under a lower power of the microscope, at which time the nuclei should appear rather densely stained, the background being only lightly stained. Each slide is removed individually from the tray, wiped on the underside with a clean cloth, and then differentiated by dropping 96 per cent alcohol (*never* acid alcohol) on it with a drop bottle or a pipette. It will be observed at once that the drops of the very thick hematoxylin solution are rolled back from the section as the 96 per cent alcohol drops on it and that, after a short time, the nuclei become more distinct and the background less distinct. The exact point at which differentiation should cease is up to the operator, but it is better, in general, since the sections are not to be counterstained, to discontinue differentiation when the nuclei are clearly defined against the background. Each slide is transferred directly to a saturated solution of lithium chloride in 70 per cent alcohol, where it passes from a pinkish color to a deep blue. If the conventional method of differentiating these stains with acid alcohol is followed, it results in a hopelessly diffuse stain. The purpose of the 96 per cent alcohol is to utilize the surface tension of the stain to hold it in the nuclei, and, if the slide is placed in acid–70 per cent alcohol, it will be found that the stain diffuses out from the nuclei, which, instead of appearing clear and sharp, appear, as it were, blurred around the edges as in an out-of-focus photograph. Differentiation by rolling back the stain with 96 per cent alcohol gives a clear sharp stain, which is as well differentiated as any of the iron alum mordant stains but which has the advantages of giving a greater transparency and also of staining the background sufficiently to render it apparent for class demonstration purposes.

The slides may remain in the saturated solution of lithium chloride in 70 per cent alcohol for as long as is required, being subsequently passed directly through the higher alcohols to xylene and mounted in balsam or some synthetic substitute for it.

This technique may, of course, be applied to accumulating and staining selected sections from any series cut from any material.

SUMMARY

1. Remove a 72-hr chick embryo from the egg and fix it as described in Example 3.
2. Wash, dehydrate, clear, remove from paper support, embed in wax, and cut a complete series of 10-μ sections throughout the whole embryo.
3. Remove from the ribbon those portions from which sections are to be mounted.
4. Flood a clean slide with diluted Mayer's albumen. Place selected sections in order, but without regard to symmetry, on fluid.
5. Using two brushes, bring the two end sections into position. Fuse the edges together in two places with a hot needle. Join the other sections in order until a ribbon has been formed. Warm to flatten sections; drain and dry.
6. Dewax dried sections and pass to 90 per cent alcohol through usual reagents.
7. Stain until nuclei darken (1 to 3 min) in Ehrlich's acid alum hematoxylin.
8. Take each slide individually and flush surface with 96 per cent alcohol from a pipette. When sufficient stain has been removed, place slide in a saturated solution of lithium carbonate in 70 per cent alcohol until blue.
9. Dehydrate, clear, and mount in balsam.

PART FOUR | Recommended Techniques

ANIMAL ORGANS

Skin and Associated Structures

Structure	Recommended source	Recommended technique
Feather bud	Feather tract on back of 13-day chicken embryo	Fix in Gilson's fluid. Stain 10-μ paraffin sections in Mallory's triple stain.
Hair	Lightly furred area of skin from rat or rabbit ear	Fix in Gilson's fluid. Stain 10-μ paraffin sections in Mallory's triple stain.
Reptilian scales	Any lizard	Fix in Bouin's fluid. Stain 10-μ paraffin sections in Patay's triple stain.
Teleost scales	Skin from posterior side	Fix in Bouin's fluid. Stain 10-μ longitudinal sections in Patay's triple stain.
Placoid scales	Laboratory dogfish	Decalcify formaldehyde-fixed skin. Stain 15-μ paraffin sections in Mallory's triple stain.
Tooth	Incisor area of jaw of 20-day cat embryo or molar area of jaw of 15-day rabbit embryo	Fix in Zenker's fluid. Decalcify. Stain 15-μ paraffin sections in Patay's triple stain.
Tongue	Rabbit	Fix in Petrunkewitsch's fluid. Stain 10-μ paraffin sections of whole organ in hematoxylin-eosin.
Lip	Newborn kitten	Fix in Zenker's fluid. Stain 10-μ paraffin sections in hematoxylin-eosin.

Nervous System and Associated Structures

Structure	Recommended source	Recommended technique
Cerebrum	Mouse	Fix in Kolmer's fluid. Stain 8-μ paraffin sections in hematoxylin-eosin.
Optic lobe	Frog	Fix whole brain in Kolmer's fluid. Stain 8-μ paraffin sections in hematoxylin-eosin.
Cerebellum	Sparrow or other small bird; rat if sparrow not available	Fix whole brain in Kolmer's fluid. Stain 8-μ paraffin sections in hematoxylin-eosin.

ANIMAL ORGANS

Nervous System and Associated Structures

Structure	Recommended source	Recommended technique
Medulla	Frog	Dissect bone from surface. Cut out whole section and fix *in situ* in Kolmer's fluid. Remove fixed medulla. Cut transverse sections through region of choroid plexus and stain in hematoxylin-eosin.
Spinal cord	Rabbit	Cut out cervical region intact and fix in Zenker's fluid. Remove cord from vertebrae after washing. Cut transverse sections and stain in hematoxylin-eosin.
Eye	Sparrow, or rat if sparrow not available	Remove eye and place in Kolmer's fluid. After 1 day, cut window in side of eye and fix 1 day more. Dehydrate slowly and carefully; embed in low-melting-point paraffin. Harden block by refrigeration. Cut 15-μ vertical sections through region of optic nerve. Stain in Patay's triple stain.
Olfactory pit	Shark embryos found in laboratory dissection specimens	Cut out olfactory area. Decalcify. Stain 10-μ paraffin sections in Mallory's triple stain.
Pineal	False chameleon (Anolis)	Cut out central area of head and fix in Zenker's fluid. Decalcify. Cut 8-μ longitudinal sections and stain in Patay's triple stain.

Digestive and Respiratory Structures

Structure	Recommended source	Recommended technique
Esophagus	Any urodele amphibian	Fix in Gilson's fluid and stain 10-μ paraffin sections in celestine blue B–eosin.
Stomach	Junction of cardiac and pyloric stomach of rat	Remove whole stomach, slit down side, rinse in saline, and fix in Gilson's fluid. Wash thoroughly, dissect out 3-mm squares from desired area, and stain 10-μ sections in celestine blue B–eosin.
Duodenum	Rat	Fix in Gilson's fluid and stain 10-μ paraffin sections in celestine blue B–eosin.

ANIMAL ORGANS

Digestive and Respiratory Structures

Structure	Recommended source	Recommended technique
Ileum	Rat	As duodenum.
Colon	Rabbit	As duodenum.
Rectum	Rabbit	Fix a short length. After thorough washing, cut through center of fecal pellet and wash out fecal matter. Stain 10-μ paraffin sections in celestine blue B–eosin.
Appendix	Rabbit	Fix in Gilson's fluid and stain 10-μ sections in celestine blue B–eosin.
Gallbladder	Rabbit	Cut out whole gallbladder with plenty of attached liver and fix in Zenker's fluid. After washing, trim closely to the bladder. Cut 10-μ sections and stain in iron hematoxylin.
Lung, simple	Frog	Dissect out lung. Gently press out as much air as possible. Inject Bouin's fixative till shape is restored. Tie off and return to fixative. After thorough washing in 70 per cent alcohol, cut 3-mm-thick transverse slices and shake out air. Stain 20-μ paraffin sections in hematoxylin-eosin.
Lung, lobular	Rabbit	Cut lung in 1-cm-thick slabs and fix in Zenker's fluid. After washing in water, transfer 5-mm cubes to 70 per cent alcohol. Leave until air has dissolved and then stain 10-μ paraffin sections in hematoxylin-eosin.

Skeletal Structures (For skeletal tissues see Connective Tissues below.)

Structure	Recommended source	Recommended technique
Bony skeleton	Small whole fish or limbs of small amphibia	Fix in slightly alkaline formaldehyde. Stain in alizarine in the manner described in text. Preserve in glycerine or mount in balsam.
Cartilagenous skeleton	Young salamander larvae or limbs of mammalian embryos	Fix in slightly acid 70 per cent alcohol. Stain in toluidine blue in manner described in text. Mount in balsam.

ANIMAL ORGANS

Skeletal Structures (For skeletal tissues see Connective Tissues below.)

Structure	Recommended source	Recommended technique
Sponge spicules	Preserved material	Cut thin slices with hand razor. Place in 5 per cent potassium hydroxide from 1 to 5 min according to density of material. Dehydrate, clear, and mount in balsam.
Arthropod skeletons	Alcohol-preserved material	Soak in 10 per cent potassium hydroxide until soft. Arrange on slide. Drop on glacial acetic acid until transparent. Rinse off acid with absolute alcohol, clear, and mount in balsam.

Glands

Structure	Recommended source	Recommended technique
Adrenal gland	Rabbit	Fix in Zenker's fluid. Stain 8-μ paraffin sections in hematoxylin-eosin.
Pituitary gland	Rabbit	Fix in Zenker's fluid. Stain 8-μ paraffin sections in hematoxylin-eosin or by one of the specialized techniques given in Gray's "Microtomist's Formulary and Guide."
Thyroid gland	Rabbit	Remove trachea with thyroid attached. Fix in Bouin's fluid. Wash thoroughly in 70 per cent alcohol, dissect off glands, and stain 10-μ paraffin sections in Patay's triple stain.
Parathyroid gland	Bullfrog	Fix in Zenker's fluid. Stain 8-μ paraffin sections in hematoxylin-eosin.
Thymus gland	Kitten	Stain 10-μ paraffin sections in celestine blue B–eosin.
Salivary gland	Rabbit or rat	Fix in Zenker's fluid. Stain 10-μ paraffin sections in hematoxylin-eosin.
Liver	Necturus or Cryptobranchus	Fix in Gilson's fluid. Stain 8-μ paraffin sections in celestine blue B–eosin.
Pancreas	Pigeon or other bird	Fix in Gilson's fluid. Stain 8-μ paraffin sections in Patay's triple stain.

ANIMAL ORGANS

Glands

Structure	Recommended source	Recommended technique
Spleen	Rabbit	Fix in Zenker's fluid. Embed in low-melting-point paraffin. Harden block by refrigeration. Cut 5-μ sections. Stain with iron hematoxylin-eosin.
Prostate gland	Rabbit	Fix in Helly's fluid. Stain 10-μ paraffin sections in hematoxylin-eosin.
Mammary gland	Rabbit, about tenth day of gestation	Cut out with adherent skin and fix in Zenker's fluid. Wash, and trim away skin to within a few millimeters of nipple. Stain 10-μ paraffin sections in hematoxylin-eosin.

Circulatory Structures

Structure	Recommended source	Recommended technique
Heart	Rabbit	Fix whole heart in Petrunkewitsch's fluid. Stain 10-μ sections of whole organ in iron hematoxylin.
Blood vessels	Sciatic artery and vein of largest mammal available	Fix in Zenker's fluid. Stain 8-μ paraffin sections in iron hematoxylin.

Renal Reproductive Structures

Structure	Recommended source	Recommended technique
Ovary, invertebrate	Earthworm	Fix a lump representing roughly the ninth to fifteenth segments in Zenker's fluid. Wash thoroughly. Open from the dorsal surface and remove ovaries from septum separating tenth and eleventh segments. Make a hematoxylin- or a carmine-stained whole mount.
Ovary, vertebrate for oocytes	Kitten	Fix in Bouin's fluid. Stain 8-μ sections in iron hematoxylin.
Ovary, vertebrate for *corpora lutea*	Rabbit	Fix in Zenker's fluid and stain 10-μ paraffin sections in Mallory's triple stain.

ANIMAL ORGANS

Renal Reproductive Structures

Structure	Recommended source	Recommended technique
Oviduct	Rabbit	Fix in Gilson's fluid. Stain 10-μ paraffin sections in hematoxylin-eosin.
Uterus	Rabbit	Fix in Gilson's fluid. Stain 10-μ paraffin sections in hematoxylin-eosin.
Placenta	Rabbit, tenth to twelfth day of gestation	Fix whole uterus in Bouin's fluid. Cut out embryo-containing swellings. Stain longitudinal 10-μ paraffin sections of whole region, including embryo, in hematoxylin-eosin.
Vagina	Kitten	Fix in Gilson's fluid. Stain 8-μ sections in hematoxylin-eosin.
Testis	Rat	Fix in Bouin's fluid. Stain 5-μ paraffin sections in iron hematoxylin.
Epididymis	Rat	Fix in Zenker's fluid. Stain 10-μ sections in iron hematoxylin.
Vas deferens	Rabbit	Fix in Gilson's fluid and stain 10-μ sections in Mallory's triple stain.
Pronephros	10-mm tadpole	Fix in Bouin's fluid. Trim away anterior end and cut transverse section of pronephric region. Stain 10-μ sections in hematoxylin-eosin.
Mesonephros	10-mm pig if available, or late larval salamander	Fix in Gilson's fluid. Stain 10-mm paraffin sections in celestine blue B–eosin.
Metanephros	Rat	Fix whole kidney in Zenker's fluid. Cut 10-μ vertical paraffin sections. Stain with celestine blue B–eosin.
Ureter	Rabbit	Fix in Gilson's fluid. Stain 8-μ paraffin sections in hematoxylin-eosin.
Penis	Rabbit	Fix in Petrunkewitsch's fluid. Stain 10-μ transverse paraffin sections in Patay's triple stain.
Bladder	Rabbit	Ligate full bladder and remove entire, to preserve shape, to Gilson's fixative. Wash thoroughly. Stain 5-μ paraffin sections from small pieces with iron hematoxylin.

ANIMAL TISSUES

Epithelial Tissues

Structure	Recommended source	Recommended technique
Squamous	Scrapings from human cheek	Make smear. When almost dry, fix in Gilson's fluid. Stain in hematoxylin.
Cubical	Skin from 10-mm tadpole	Fix in Bouin's fluid. Stain 8-μ paraffin sections in hematoxylin-eosin.
Columnar	Rabbit intestine	See Animal Organs above.
Ciliated	Frog pharynx or skin from roof of frog's mouth	Fix in Gilson's fluid. Stain 8-μ paraffin sections in hematoxylin-eosin.
Stratified	Rabbit bladder	See Animal Organs above.

Connective Tissues

Structure	Recommended source	Recommended technique
White fibrous	Ligament of frog gastronemius	Fix in Petrunkewitsch's fluid. Stain 8-μ longitudinal sections in Mallory's triple stain.
Yellow elastic	Patagial tendon of pigeon	Fix in Zenker's fluid. Stain 8-μ longitudinal sections in iron hematoxylin or use one of the special techniques given in Gray's "Microtomist's Formulary and Guide."
Areolar	Mesentery of rabbit	Half fill small vials with Bouin's fluid. Tie mesentery as a cap over vial and invert the vial in container of Bouin's fluid. Cut out disk of hardened mesentery and make hematoxylin-stained wholemount.
Cartilage, hyaline	Aretenoid cartilage of rabbit	Fix in Bouin's fluid. Stain 10-μ paraffin sections in hematoxylin.
Cartilage, fibrous	Pubic symphysis of rabbit	Cut out symphysis with bone forceps. Fix in Zenker's fluid. Decalcify. Cut 10-μ transverse paraffin sections and stain in Mallory's triple stain.
Cartilage, elastic	Rabbit ear	Fix 5-mm squares from lower third of center of rabbit ear in Zenker's fluid. Stain 8-μ sections in iron hematoxylin or use one of special methods given in Gray's "Microtomist's Formulary and Guide."

ANIMAL TISSUES

Connective Tissues

Structure	Recommended source	Recommended technique
Muscle, striped	Muscle from crayfish claw	Fix in Petrunkewitsch's fluid. Stain 10-μ longitudinal paraffin sections in iron hematoxylin.
Muscle, cardiac	Any vertebrate heart	Fix in Petrunkewitsch's fluid. Stain 10-μ longitudinal sections in iron hematoxylin.
Muscle, smooth	Rabbit intestine	Fix ¼-in. lengths in Gilson's fluid. Cut 10-μ tangential sections through region of longitudinal muscle. Stain in iron hematoxylin.
Bone, Haversian	Rabbit femur	Fix femur in Zenker's fluid. Saw ¼-in. sections from central portion and decalcify. Cut 15-μ transverse paraffin sections and mount on slide. Dissolve paraffin in xylol, allow specimen to dry, and then mount in balsam.
Bone, laminar	Rabbit femur	Prepare as in last example but cut longitudinal tangential sections from surface layer. Mount, unstained, in balsam.
Blood	Any vertebrate	Make smear and stain with Wright's stain as described in text.
Pigment cells	Liver of Necturus or Cryptobranchus	Fix in Zenker's fluid. Mount unstained 10-μ sections in balsam.

Nervous and Sensory Tissues

Structure	Recommended source	Recommended technique
Giant motor cells	Rabbit spinal cord	Make smears from ventral horn of cord. When almost dry, fix in Gilson's fluid. Stain in hematoxylin.
Pyramidal, neuroglia, and Purkinje cells		Can be properly demonstrated only by specialized metal staining techniques given in Gray's "Microtomist's Formulary and Guide."
Medullated fibers	Sciatic nerve of rabbit	Remove about a ½-in. length of sciatic nerve to a drop of saline on a slide. Tease out one end as finely as possible with needles. Add a drop of

ANIMAL TISSUES

Nervous and Sensory Tissues

Structure	Recommended source	Recommended technique
		osmic acid. Leave 5 min. Wash off osmic acid with saline. Dehydrate with absolute alcohol dropped from pipette. Clear with terpineol and complete wholemount with balsam.
Motor nerve ends	Superior oblique muscle of rabbit orbit	Stain by lemon juice–gold chloride method of Ranvier given in Gray's "Microtomist's Formulary and Guide."
Retina	Rabbit eye	Cut off back half of rabbit eye and fix in Kolmer's fluid. Stain 5-μ paraffin sections of small pieces in iron hematoxylin.

ANIMAL CYTOLOGY

Nuclei

Structure	Recommended source	Recommended technique
Resting nuclei	Necturus liver	Fix in Bouin's fluid. Stain 10-μ paraffin sections in celestine blue B. Do not counterstain.
Mitosis, many chromosomes	Skin of 10-mm tadpole tail	Fix whole tadpole in Bouin's fluid. Wash thoroughly in 70 per cent alcohol and then transfer to water. Cut flap of skin in posterior abdominal wall and strip this backward to remove surface skin from tail. Stain in Ehrlich's hematoxylin and make wholemount in balsam.
Mitosis, few chromosomes	Ascaris	Slit open Ascaris and fix in Carnoy and Lebrun's fixative. Wash in absolute alcohol and transfer to 70 per cent alcohol. Cut oviduct into ¾-in. lengths and tie in loose ¼-in. bundles, assorting the regions. Cut 10-μ transverse sections of bundles and mount ribbons on slide. Stain in iron hematoxylin.
Meiosis	Ascaris	As in the last example, but using bundles of ovary.

ANIMAL CYTOLOGY

Nuclei

Structure	Recommended source	Recommended technique
Giant chromosomes	Salivary gland of Drosophila	See text for complete details of procedure.

Cytoplasmic Organelles

Structure	Recommended source	Recommended technique
Mitochondria	Liver	Fix in Zenker's fluid. Stain 5-μ paraffin sections with hot Ziehl's carbolmagenta. This method, though unconventional, is often successful. For other methods, see Gray's "Microtomist's Formulary and Guide."
Golgi bodies	Earthworm ovary	Fix in 2 per cent osmic acid overnight. Mount 5-μ paraffin sections without further staining.
Nissl bodies	Rabbit	Make smear of ventral horn of spinal cord as described under Nervous Tissues above. Stain in strong methylene blue solution. This method, though unconventional, is often successful. For detailed description of other methods, see Gray's Microtomist's Formulary and Guide."
Myofibrils	Stentor	Fix Stentor in hot Helly's fluid. Make 5-μ longitudinal paraffin sections and stain in Mallory's triple stain.

PLANT ORGANS

Stem

Structure	Recommended source	Recommended technique
Monocot	Corn	Fix in Bles's fluid. Stain 10-μ paraffin sections in Johansen's quadruple stain.
Dicot	Willow	Fix young shoots in spring in Bles's fluid. Stain 10-μ sections in Johansen's quadruple stain.

PLANT ORGANS

Stem

Structure	Recommended source	Recommended technique
Fern	Dennstaedtia	Fix in Bles's fluid. Stain 10-μ sections in safranin–fast green.
Lycopod	Lycopodium or Selaginella	Fix in Bles's fluid. Stain 10-μ paraffin sections in safranin–fast green.
Horsetail	Equisetum	Fix young stems in Bles's fluid. Stain 10-μ transverse sections in safranin–fast green.
Quillwort	Isoetes	Fix ½-in. lengths in Bles's fluid. Cut in 2-mm lengths and leave in 70 per cent alcohol until air is dissolved. Stain 10-μ paraffin sections in safranin–fast green.
Xerophyte	Alloe	Fix in Bles's fluid. Stain 10-μ paraffin sections in Johansen's quadruple stain.
Succulent	Sedum	Fix in Bles's fluid. Stain 10-μ paraffin sections in Johansen's quadruple stain.
Aquatic	Typha	Fix in Bles's fluid. Stain 10-μ paraffin sections in safranin–fast green.
Tuber	Potato	Fix 5-mm cubes in Bles's fluid. Stain 10-μ paraffin sections in safranin–light green.

Root

Structure	Recommended source	Recommended technique
Dicot	Ranunculus	See text for full description.
Epiphyte	Cattleya	Fix in Bles's fluid. Stain 10-μ transverse sections in Chamberlain's double stain.
Origin of secondary root	Vicia seedlings	Fix in Bles's fluid. Stain 10-μ vertical sections in safranin–fast green.

Leaf

Structure	Recommended source	Recommended technique
Dicot	Ligustrum	Fix small whole leaves in Bles's fluid. Leave 3-mm squares in 70 per cent alcohol until air is dissolved. Stain 10-μ transverse sections in safranin–fast green.

PLANT ORGANS

Leaf

Structure	Recommended source	Recommended technique
Monocot	Iris	Fix in Bles's fluid. Stain 10-μ paraffin sections in safranin–fast green.
Aquatic	Elodia	Fix in Bles's fluid. Stain 10-μ paraffin sections in safranin–fast green.
Succulent	Sedum	Fix in Bles's fluid. Stain 10-μ paraffin sections in Johansen's quadruple stain.
Xerophytic	Ammophila	Fix in Bles's fluid. Stain 10-μ paraffin sections in Johansen's quadruple stain.
Development	Syringa bud	Peel outer leaves from small dormant bud. Fix in Bles's fluid. Stain 10-μ vertical paraffin sections in hematoxylin.

Flower

Structure	Recommended source	Recommended technique
Bud	Ranunculus	Fix young buds in Navashin's fluid. Stain 10-μ paraffin sections in hematoxylin.
Stamen	Lily	Remove stamens to Navashin's fluid. Select straight specimens for vertical 8-μ sections. Stain in hematoxylin.
Pistil	Lilium	Fix whole pistil in Navashin's fluid. Stain 10-μ vertical sections in Johansen's quadruple stain.
Ovary	Lilium	Fix in Navashin's fluid. Stain 10-μ transverse sections in safranin–fast green.

PLANT TISSUES

Structure	Recommended source	Recommended technique
Epidermis	Tradescantia	Strip epidermis from alcohol-fixed specimens and make wholemount stained in hematoxylin.
Cortex	Any plant stem	See Stem above.
Pith, in stem	Aristolochia	Fix in alcohol. Stain 10-μ transverse sections in hematoxylin.

PLANT TISSUES

Structure	Recommended source	Recommended technique
Pith, in root	Smilax	Fix in alcohol. Stain 10-μ transverse sections in hematoxylin.
Phellogen	Betula	Fix young stem in Navashin's fluid. Stain longitudinal 10-μ sections in hematoxylin.
Phellem	Philodendron	Fix in Navashin's fluid. Stain 10-μ longitudinal sections in safranin.
Phloem, primary	Pteridium rhizome	Fix in Bles's fluid. Stain 10-μ transverse sections in safranin–fast green.
Phloem, secondary	Magnolia	Fix very young twigs in Bles's fluid. Stain transverse sections in safranin–fast green.
Xylem, primary	Pteridium rhizome	Fix in Navashin's fluid. Stain 10-μ transverse sections in Chamberlain's double stain.
Vascular cylinder, protostele	Psilatum (aerial stem)	Fix in Bles's fluid. Stain 10-μ transverse sections in safranin.
Ectophloic siphonostele	Ophioglossum	Fix in Navashin's fluid. Stain in Chamberlain's double stain.
Amphiphloic siphonostele	Pellaea	Fix in Navashin's fluid. Stain 10-μ sections in Chamberlain's double stain.
Ectophloic dictyostele	Osmunda	Fix in Navashin's fluid. Stain 10-μ sections in Chamberlain's double stain.
Amphiphloic dictyostele	Polystichum	Fix in Navashin's fluid. Stain 10-μ sections in Chamberlain's double stain.
Eustele	Helianthus	Fix young stems in Navashin's fluid. Stain in Johansen's quadruple stain.
Atactostele	Smilax	Fix young stems in Bles's fluid. Stain 10-μ transverse sections in safranin.
Xylem, secondary in dicot	Quercus	Cut tangential sections of boiled wood block as thin as possible. Stain in safranin.
Xylem, secondary in monocot	Palm	Cut tangential sections of boiled wood block. Stain in safranin.
Xylem, secondary in fern	Botrychium	Fix in Bles's fluid. Stain 10-μ transverse sections in safranin.
Xylem, secondary in root	Daucos	Fix in Bles's fluid. Stain 10-μ transverse sections in safranin.

PLANT TISSUES

Structure	Recommended source	Recommended technique
Parenchyma	Solanum	Fix in Bles's fluid. Stain transverse sections in fast green.
Collenchyma	Pelargonium	Fix in Bles's fluid. Stain 10-μ transverse sections in fast green.
Sclerenchyma	Tilia	Fix very young stems in Navashin's fluid. Stain 10-μ longitudinal sections in light green.
Vascular cambium	Pinus	Cut longitudinal sections of boiled wood block. Stain in safranin.
Haustoria	Dodder	Fix whole parasitized stems in Navashin's fluid. Cut 10-μ transverse sections through region of attachment and stain in Johansen's quadruple stain.
Mycorrhiza	Roots of hickory	Fix in Navashin's fluid. Stain 10-μ longitudinal sections in safranin. Counterstain with Smith's picro–spirit blue.

PLANT CYTOLOGY

Structure	Recommended source	Recommended technique
Resin ducts	Pinus stem	Fix in alcohol. Cut 10-μ longitudinal sections. Stain in safranin.
Laticifers	Euphorbia	Fix in alcohol. Cut 20-μ longitudinal frozen sections. Stain in Sudan IV and mount in glycerine jelly.
Abscission layer	Coleus leaf	Fix in Bles's fluid. Stain 10-μ sections in hematoxylin.
Hypodermis	Ficus leaf	Fix in Bles's fluid. Stain 10-μ sections in hematoxylin.
Secretory cells, digestive	Drosera leaf	Fix whole leaf. Stain 10-μ paraffin sections in hematoxylin.
Secretory cells, nectary	Narcissus petal	Fix in Navashin's fluid. Cut 10-μ paraffin sections longitudinally through base of petal. Stain in hematoxylin.
Trichomes	Nerium leaf	Fix in alcohol. Strip epidermis and make unstained mount in either glycerine jelly or balsam.

PLANT CYTOLOGY

Structure	Recommended source	Recommended technique
Vessel elements	Rhubarb	Boil short length until soft. Squash small pieces on slide and mount in glycerine jelly.
Tracheid	Pinus	Cut longitudinal section of boiled wood block. Stain in safranin.
Wood fibers	Quercus	Macerate fragments by Harlow's technique. Shake with glass beads to separate. Wash, centrifuge, stain in safranin, wash, centrifuge, and mount in balsam.
Sieve cells	Pinus	As for wood fibers.
Sieve-tube elements	Tilia	As for wood fibers.
Companion cells	Squash stem	Fix in Bles's fluid. Stain 10-μ longitudinal sections of phloem region in safranin.
Phloem fibers	Flax stem	Fix in Navashin's fluid. Stain 8-μ longitudinal sections in hematoxylin.
Cambium, cork	Prunus	Fix in Bles's fluid. Stain 10-μ longitudinal sections in hematoxylin.
Cambium, vascular	Pinus	Cut 10-μ longitudinal sections from boiled block of wood. Stain in safranin.
Lenticels	Sambucus	Strip bark and fix in Navashin's fluid. Cut 10-μ longitudinal sections and stain in hematoxylin.
Pits	Applewood	Fix in Navashin's fluid. Cut 10-μ longitudinal sections of young twig and stain in safranin.
Sclereids	Camellia	Fix leaves in alcohol. Stain 5-mm squares in safranin, differentiate fully, and make balsam wholemount.
Crystals, raphides	Jerusalem artichoke	Fix tuber in alcohol. Clear hand sections in glycerine. Mount in glycerine jelly.
Crystals, druses	Tilia	Fix young stems in alcohol. Cut hand sections through stem cortex. Clear in glycerine. Mount in glycerine jelly.
Crystals, stone cells	Kieffer pears	Preserve in alcohol. Cut hand sections. Clear in glycerine. Mount in glycerine jelly.

PLANT CYTOLOGY

Structure	Recommended source	Recommended technique
Crystals, prismatic	Rhubarb leaf	Fix in alcohol. Clear in xylol. Mount in balsam.
Starch grains	Potato	Fix in alcohol. Cut 8-μ paraffin sections at right angle to skin. Stain in safranin.
Plasmodesmata	Date pit	Boil pit until softened. Cut tangential sections. Mount unstained in balsam.
Megasporogenesis	Lily ovary	Squash in aceto-orcein. Render permanent by method given in text for Drosophila chromosome.
Microsporogenesis	Tradescantia anthers	Squash in aceto-orcein. Render permanent by method given in text for Drosophila chromosome.
Mitosis	Onion root tip	Fix in Navashin's fluid. Stain 10-μ longitudinal sections in iron alum.
Meiosis	Fern sporangia	Squash in aceto-orcein. Make permanent by technique given in text for Drosophila chromosome.
Chromosomes, in squash	Vicia root tip	Squash in aceto-orcein. Make permanent by technique given in text for Drosophila chromosome.
Chromosomes, in section	Onion root tip	Fix in Navashin's fluid. Cut 8-μ longitudinal sections. Stain in celestine blue B.

SUGGESTED ADDITIONAL SOURCES OF INFORMATION

Baker, J. R.: "Cytological Techniques," 4th ed., New York, John Wiley & Sons, Inc., 1960.

Carleton, H. M., and R. A. B. Drury: "Histological Technique," 3d ed., Fair Lawn, N.J., Oxford University Press, 1957.

Cowdry, E. V.: "Laboratory Technique in Biology and Medicine," 3d ed., Baltimore, The Williams & Wilkins Company, 1952.

Darlington, C. D., and L. F. LaCour: "The Handling of Chromosomes," 2d ed., London, George Allen & Unwin, Ltd., 1948.

Gray, F., and P. Gray: "Annotated Bibliography of Works in Latin Alphabet Languages on Biological Microtechnique," Dubuque, Iowa, W. C. Brown Company, 1956.

Gray, P.: "The Microtomist's Formulary and Guide," New York, McGraw-Hill Book Company, 1954.

Guyer, M. F.: "Animal Micrology," 5th ed., Chicago, The University of Chicago Press, 1953.

Humason, G. L.: "Animal Tissue Techniques," San Francisco, W. H. Freeman and Company, 1962.

Jones, R. M.: "Microscopical Technique," 3d ed., New York, Paul B. Hoeber, Inc., 1950.

Johansen, D. A.: "Plant Microtechnique," New York, McGraw-Hill Book Company, 1940.

INDEX

Abbe substage condenser, 19, 36
Aberrations of lenses, 4–8
   correction of, 7
Abscission layer, method for, 284
Acetic acid, in fixative, 88–91
   as ingredient of stain, 115
   in staining solution, 102
Aceto-orcein, 106
Achromatic, explanation of, 14–15
Achromatic objective, 44
Achromatic substage condenser, 19
Acid alcohol, 101, 103, 105
Acid fixer, photographic, 62
Acid fuchsin, applied, to animal tissues, 110
   to plant tissues, 235
   in complex stain, 111
Acids (see name, e.g., Picric acid)
Adhesive, for frozen sections, 189
   for paraffin sections, 173–174
Adrenal, method for, 274
Agar, for attaching frozen sections, 189
   for temporary wholemounts, 129
Alcohol, dehydrating agents for, 123, 239
   in fixative, 89, 90
   for preserving plant tissues, 233
   recommended series of, 118–119, 134
   for storing wax blocks, 161
   substitutes for, 119
Alcoholic hematoxylin, 101
Alizarin red S, 116
Allen, R. M., 73
Allen's fixative, 91

Allen's fixative, applied to rat testis, 250–251
Alloe, method for stem, 281
Alum as mordant, 98
American Optical Company, 32, 43
Ammonium acetate for fixative removal, 93
Ammonium alum in staining solutions, 102, 103
Ammonium oxalate as ingredient of stain, 188
Ammophila, method for leaf, 282
Amphioxus, preparation of section, 258–260
Amplitude of light, 22
Anderson's medium, 186
Angular aperture, 9, 10
Aniline blue, staining with, 111
Animal cytology, methods for, 279–280
Animal organs, methods for, 271–276
Animal tissues, methods for, 277–279
Annelida, wholemounts, 132
Annular plate, centering, 53
Annular stop, 28
Anolis, method for pineal, 272
Antiformin, 228
Apochromatic, explanation of, 15
Apochromatic objective, 44
Appendix, method for, 273
Aquatic leaf, method for, 282
Aquatic stem, method for, 281
Ascaris, methods for chromosomes, 279
Areolar tissue, method for, 277
Aristolochia, section of stem, 72, 245–249
Arthropods, fixative for, 90

289

Arthropods, mountant for, 125
  skeletons, method for, 274
  softening, 94, 198
Auxochrome, 98
Azo, 97
Azure II, 114
  for staining sections, 262

Bacteria in sections, 114, 261–263
  nucleus in, 70
  smear of, 222–229
Baker, J. R., 96, 117, 286
Balsam (*see* Canada balsam)
Basic fuchsin (*see* Magenta)
Bausch and Lomb microscope, 45, 68
Bayberry wax in embedding media, 152
Beeswax in embedding media, 152
Bennett, A. H., 30
Benzene, for clearing tissues, 121–122, 239–240
  unsuitability for wholemounts, 135
Berlese's funnel, 197–198
Berlese's mountant, 125
  used for mite, 198–199
Birchon, D., 54
Bladder, method for, 276
Blood, method for, 278
Blood smear, 113–114, 140–143, 219–221
Blood stains, explanation of, 98
Blood vessels, method for, 275
Bone, decalcifying, 93–94
  methods for, 278
  staining, 113, 116
Bony skeleton, method for, 273
Borax as ingredient of stain, 104, 114
Bouin's fixative, 90
  softening tissues fixed in, 258
  washing from tissues, 93
Bright phase, 28
Brightness, apparent, 26–27
Bryozoa, 85
  mounting, 200–205
  narcotizing, 95
  wholemounts, 133
Bud, method for, 282
Butyl alcohol (*see* Tert-butyl alcohol)

Cabinets for storing slides, 193–194
Cambium, method for, 285
Cameras, photomicrographic, 66–69
Campeche wood, 99
Camphor as ingredient of mountant, 126
Camsal, 126
Canada balsam, finishing mounts in, 139
  preparing "dry," 127
  technique of mounting in, 136–139
  types of, 126–127
Carbol fuchsin (*see* Carbolmagenta)
Carbolmagenta, 108
  applied to bacteria, 228
Carbolxylene, 121
Carboxymethyl cellulose for temporary wholemounts, 129
Cardiac muscle, method for, 278
Carleton, H. M., 286
Carmalum, 105
Carmine, for staining, liver fluke, 214–215
  Pectinatella, 200–205
  stains, 104–106
Carnoy and Lebrun's fixative, 90
Cartilage, methods for, 277
  staining, 112, 113, 116
Cartilagenous skeleton, method for, 273
Cat (*see* Kitten)
Cattleya, method for root, 281
Celestine blue B, for bacterial nucleus, 70
  for intestine, 242
  stain, 106–107
  for tongue, 256
  (*See also* Gray's celestine blue B)
Cell for wholemounts, 130
Cellosolve, as dehydrant, 119–120
  for differentiating stains, 115
  as ingredient of stain, 107
  as narcotic, 95
Centering mechanical stage, 51
Central stop, 29
Cerebellum, method for, 271
Chamberlain's stain, applied to root, 233–236
Chick embryo, method for feather bud, 271
  section of, 264–267
  wholemounts of, 206–211
Chitin, softening, 94, 198

Chloral hydrate, as ingredient of mountant, 125
  as narcotic, 95, 132
  for softening chitin, 94
Chlorine for macerating wood, 94
Chloroform, as clearing agent, 121
  in fixative, 90
  as narcotic, 133
  as solvent for balsam, 127
Chromatic aberrations, 6
Chrome-acetic fixative for Bryozoa, 133
Chrome-alum for mounting sections, 175
Chromic acid fixatives, 88–89, 91
Chromic oxide (*see* Chromic acid fixatives)
Chromophore, 97
Chromosomes, fixation of, 89–90
  method for, 279–280
  plant, methods for, 286
  in rat testis, 250–254
  salivary, 230–232
  stains for, 106–107
Ciliated epithelium, method for, 277
Circulatory structures, methods for, 275
Cladocera, 198
Cleaning, coverslips, 243
  prepared slides, 191–192
  slides before use, 140–141, 241
Clearing agents, 120–123
  flotation technique, 122, 204
  for wholemounts, 135
Clove oil, for clearing, 121
  in staining technique, 115
  for wholemounts, 135
Coagulant fixatives, 86
Cochineal, 99
Coddington lens, 7
Coelenterata, narcotizing, 96
  wholemounts, 132
Collembola, 198
Collenchyma, method for, 284
Colon, method for, 273
Color filters, 21, 23–34
  of light, 23
Columnar epithelium, method for, 277
Companion cells, method for, 285
Compensating ocular, 15
Complex contrast stains, 111–115

Condenser (*see* Field condenser; Substage condensers)
Conn, H. J., 117
Connective tissues, methods for, 112, 113, 277–278
Containers, 81–82
Contrast in photomicrography, 58–59, 72
Contrast stains, 109
Coplin jar, 83
Copper sulfate for dehydrating, 123
Cork, methods for, 285
Corn, method for stem, 280
Corpora lutea, methods for, 275
Cortex, method for, 282
Coverslips, cleaning, 243
  placing, on sections, 182
  on wholemounts, 136–138
  specifications for, 80
  supporting over thick objects, 130
Cowdry, E. V., 286
"Craf" fixatives, 89
Crayfish, method for muscle, 278
Cryptobranchus, method for liver, 274
Crystal violet, applied to bacteria, 222–224
  Lillie's, 108
Crystals in plant tissues, 285–286
Cubical epithelium, method for, 277
Cupric nitrate in fixative, 91
Cut film, for photomicrography, 70–71
  processing, 64
Cutting, freehand sections, 146–149
  frozen sections, 187–189
  paraffin sections, 165, 170–185
Cutting facet, 164–165
Cytology, animal, methods for, 279–280
  plant, methods for, 284–286
Cytoplasm, staining, 98

Dark field microscopy, 29
Dark phase, 28
Dark slide, 71
Darlington, C. D., 286
Decalcification, 93–94
Dehydration, with cellosolve, 119
  with dioxane, 119
  with ethyl alcohol, 118–119

Dehydration, Johansen's mixture for, 115
  of plant tissue, 246–248
  principles of, 118
  technique of, 134–135, 238–239
Delafield's hematoxylin, 102
  for chick embryo, 210–211
  for embryo wholemounts, 133
  in triple stain, 112
Dennstaedtia, method for stem, 281
Development, of papers, 64–66
  of photographic negatives, 61–64
  theory of, 56
Deviated wave, 22, 24–29
Dextrin in embedding medium, 186
Dextrose in mountants, 125
Diagrams of slide preparation, 78–79
Diaphragm, iris (see Iris diaphragm)
Dichromate fixatives, 87–88
Dicot, leaf, method for, 281
  root, method for, 281
  stem, method for, 280
Differential staining, theory of, 97–99
Digestive system, methods for, 272–273
Dioxane, as dehydrant, 119, 150
  in mountant, 126
Diplococci in section, 261–263
Direct hematoxylin stains, 101–103
Direct image projection, 66–67
Direct staining, 98–99
Dispersion, 5
Distomum (see Liver fluke)
Dobell's hematoxylin, 101
Double contrast stains, 109
Doublet lens, 7
Drierite, 123
Drosophila, method for chromosomes, 230–233, 280
Drury, R. A. B., 286
Druses, method for, 285
Duodenum, method for, 272
Dyes, nature of, 97
  types of, 98

Earthworm, method for ovary, 275
E.F. (equivalent focus), 13
Egg, incubation of, 206

Egg albumen, as ribbon adhesive, 173
  for mounting, ameba, 132
  sections, 153
  removing embryo from, 207–209
Ehrlich's hematoxylin, 102
  for chick sections, 266
Elastic tissues, methods for, 277
Elder pith, 147
Elodia, method for leaf, 282
Embedding, technique of, 159–162
Embedding boxes, 156–158
Embedding ovens, 154–155
Embedding waxes, 152
  for frozen sections, 186
Embryo, skeletons of, 273
Embryo chick (see Chick embryo)
Embryological watch glasses, 82
Eosin, in blood stain, 113
  used for sections of intestine, 243
Eosin Y, 109
Epidermis, plant, method for, 282
Epididymis, method for, 276
Epiphyte root, method for, 281
Epithelia, methods for, 277
Equisetum, method for stem, 281
Equivalent focus (E.F.), 13
Esophagus, method for, 272
Essential oils, 120–121
Ether, in fixative, 91
  as narcotic, 133
Ethyl alcohol (see Alcohol)
Ethyl eosin, 109
Eucaine hydrochloride as narcotic, 95
Euparal, 126
Exposure, 72
External form, preservation of, 85, 86
Eye, fixation of, 91
  method for, 272
Eyepiece (see Ocular)

Farrants's mountant, 124
Fast green FCF, 109, 115
  used for Aristolochia section, 248
Fats, staining, 116
Feather, method for, 271
Fern, stem, method for, 281

Fibers, plant, methods for, 285
Fibrous tissues, methods for, 277
Field condenser, 36
  centering, 48
Field iris, 38
Film processing, 63–64
Filter, passage of light through, 23–24
Filters for photomicrography, 71–72
Fixation, Ameba, 132
  Amphioxus, 258
  annelida, 132
  Aristolochia stem, 245
  bryozoa, 133, 201
  chick embryo, 209–210
  chromosomes, 89–90
  coelenterata, 132
  eyes, 91
  liver fluke, 212–214
  marine forms, 89
  Pectinatella, 201
  photographic materials, 56
  plant tissues, 89, 245
  protozoa, 132
  purpose of, 85
  rat testis, 250–251
Fixative formulas, alkaline, 87
  mixtures, action of, 86–87
    defined, 85
  removal from tissues, 92–93
  (See also name, e.g., Zenker's fixative, or ingredient, e.g., Picric acid)
Flat worms, wholemounts of, 132
Flattening, liver flukes, 212–214
  paraffin sections, 175–177
  wholemounts, 201–202
Flower bud, method for, 282
Flowers, methods for, 282
Fluorite, 8
Formaldehyde, in fixative, 88–91
  for fixing bryozoa, 201
Freehand sections, general description, 146–149
  microtome for, 146
  supporting substances for, 147
Freezing to preserve squashes, 144, 232
Freshman microscope (*see* Microscope, freshman)

Frog, method for, lung, 273
  medulla, 272
  optic lobe, 271
  parathyroid, 274
Frog intestine, sections of, 237–244
Frozen sections, attaching to slide, 189–190
  cutting, 187–189
  microtome for, 185–186
  mounting, 189–190

Gage, S. H., 30
Gallbladder, method for, 273
Gamasid mites, 198
Gates's fixatives, 89
Gelatin for mounting sections, 174
Gerhardt's fixative, 210
Giant chromosomes, method for, 280
Gilson's fixative, 89
  for chick embryo, 209
  for liver fluke, 213
Glands, method for, 274–275
Glare, effect of, 18–19
Glycerin, in mountants, 125
  in staining mixture, 100, 102
Glycerin jelly, 198
Gold size, 130
Golgi bodies, method for, 280
Gore, 94
Gram's iodine, 92
  applied to bacterial smear, 225–226
Gravis's adhesive, 189
  used for mites, 198
Gray, F., 286
Gray, P., 96, 117, 286, 287
Gray and Wess's medium, 125
Gray's celestine blue B, 106
  double-contrast stain, 110
  for frog intestine, 242
  narcotic, 95
  for rat tongue, 256
Grenacher's carmine, 104
  for invertebrates, 133
  for Pectinatella, 200–205
Gum acacia as mounting medium, 124–125
Gum mountants, 124–126
  finishing slides in, 130

Gum sandarac as mountant, 126
Gurr, E., 117
Guyer, M. F., 287

Hair, method for, 271
Hance's rubber paraffin, 152
Hanley's solution, 95
Hard materials, fixative for, 90
 softening, 93–95
Hardening photographic negatives, 62
Harlow's macerating solution, 94–95
Haug's solution, 93
Haustoria, method for, 284
Heart, method for, 275
Heat as immobilizing agent, 86
Heidenhain's fixative, 88
Heidenhain's hematoxylin, 99
 for rat testis, 252–253
Helly's fixative, 88
 for frog intestine, 237
Hematein, staining with, 101
Hematoxylin, for chick sections, 266
 for chick wholemount, 210–211
 for monocystis, 217
 staining, direct, 103–104
 stains, 99–104, 266
  bluing, 102–103
 for testis, 252–253
Horsetail stem, method for, 281
Hucker's crystal violet, 108
Humason, G. L., 287
Huygenian ocular, 15
Hyaline cartilage, methods for, 277
Hydra, squash of, 144
Hydrochloric acid, for differentiating, 101
 as ingredient of stain, 116
Hydrozoa, 86
"Hypo," 62
Hypodermis, method for, 284

Ileum, method for, 273
Illuminant, built-in, 41
 for freshman microscope, 32
 for medical microscope, 36, 41

Illuminant, requirements of, 20
 for research microscope, 46–47
 separate, 36–40
 setting up, 41
Illumination, Kohler, 37–38
Image, direct projection of, 66–67
 formation of, 4, 6–7
 how seen, 22–30
 reflex projection of, 67, 69
 in relation to background, 26
 split beam, 67, 69
Immersion lenses, reason for, 12
 setting up, 40, 50
Immersion oil, use with condenser, 11
Immobilizing agents, 86
Incubation of hen's eggs, 206
Indamine, 97
Index of refraction, 8
 as affecting resolution, 10–12
 of mounting media, 124
Indirect staining, 98–99
 carmine, 104–105
 hematoxylin, 103
Invertebrates, narcotizing, 131–133
 staining, 133
Iodine, Lugol's, 92
Iodine green, applied to plant tissues, 235
Ionizing groups, 98
"Ipso" camera, 68
Iris, method for leaf, 282
Iris diaphragm, 7
 field, 38
 function of, 16
 as light source, 37
 substage, 16
Iron alum, as ingredient of stain, 107
 as mordant, 100–101
Iron hematoxylin, for plant wholemounts, 134
 stains, 100–101
  technique of differentiating, 252–253
Isoetes, method for stem, 281

Johansen, D. A., 190
Johansen's quadruple stain, 114–115
 for plant stem, 72, 245–249
Johansen's safranin, 107

Jones, R. M., 287
Jurray's mixture, 94

Kitten, method for, ovary, 275
  thymus, 274
  tongue, 271
  vagina, 276
Knives, microtome, sharpening, 165–168
Kohler illumination, 37–38
Kolmer's fixative, 91

Labeling slides, 139, 192–193
LaCour, L.F., 286
LaCour's aceto-orcein, 106
  for salivary-gland chromosomes, 230–232
Lactic acid in mountants, 125
Lactophenol for swelling dried plant specimens, 233
Lamp (*see* Light sources)
Latent image, 56
Laticifers, method for, 284
Lavdowsky's fixative, 87
  for plant stem, 245
Leaf, methods for, 281–282
  structures in, methods for, 284
Lebrun (*see* Carnoy and Lebrun's fixative)
Leeches, wholemounts, 133
Leeuwenhoek, Anton van, 3
Leica, 68
Lenoir's fluid, 93
Lens, aberrations of, 4–8
  Coddington, 7
  doublet, 7
  meniscus, 7
  microscope (*see* Objective; Ocular; etc.)
Lenticels, method for, 285
Leucocyte staining, 219–220
Ligament, method for, 277
Light, absorption of, 23–24
  color of, 23
  dispersion, 5
  nature of, 22
  phase of, 26–27
  refraction, 5
Light green, 112

Light sources, centering, 20–22, 49
Ligustrum, method for leaf, 281
Lillie's crystal violet, 108
  for bacterial smear, 222–224
Lily, method for flower, 282
Lip, methods for, 271
Lithium carbonate for removal of picric acid, 92, 250–251
Liver, method for, 274
Liver fluke, preparation of wholemount, 212–215
LoBianco's fixative, 88
Lugol's iodine, 92
Lungs, method for, 273
Lycopod stem, method for, 281

Macerating wood, 94–95
McNamara, Murphy, and Gore's solution, 94
Magenta, 108
  for bacteria, 228
Magnesium sulfate as narcotic, 95, 133
Magnification, 12–14
Mallory's triple stains, 111, 114
  for Amphioxus, 259
  for Diplococci in liver, 262
Mammary gland, method for, 275
Maxwell's embedding wax, 152
Mayer's albumen, 173
Mayer's carmalum, 105
  for liver fluke, 214–215
Mechanical stage, 51
Medical microscope (*see* Microscope, medical)
Medulla, method for, 272
Medullated fibers, method for, 278
Megasporogenesis, method for, 286
Meiosis, 153, 279, 286
Meniscus lens, 7
Menthol, as narcotic, 95, 133
  for narcotizing, bryozoa, 201
  liver fluke, 212
Mercuric chloride fixatives, 87–90, 132
  in decalcifier, 94
  removal from tissues, 92
Mesentery, method for, 277
Mesonephros, method for, 276

Metanephros, method for, 276
Methanol in staining solution, 102
Methods, abbreviated, for specific structures, organs, or tissues (*see* name of structure, organ, etc.)
Methyl blue, applied as Mallory's stain, 259
Methyl violet, 115
   for Aristolochia, 248
Methylene blue, 114
   for blood smear, 144
   for sections, 262
Micron, definition, 147
Microscope, diagram of, 4
   for Kohler illumination, 38
   freshman, N.A., 11
     objectives for, 32
     setting up, 32–34
     specifications, 31–32
   lens systems, 25, 28
   lenses (*see* Objective; Ocular; etc.)
   medical, 33–41
     illuminant for, 36, 41
     objectives for, 35
     setting up, 36–42
       with built-in illuminator, 41
     specifications, 34
   phase-contrast, 27–28
     setting up, 51–54
   research, 43–51
     illuminating system, 43, 47
     objectives for, 44–45
     setting up, 47–51
     specifications, 42
     substage, 44–45
   slide (*see* Section; Slides; Smears; Wholemounts; etc.)
   stage, 43
   trinocular, 67–68
Microsporogenesis, method for, 286
Microstar camera, 68
Microstomum, wholemount, 132
Microtome, freehand, 146
   freezing, 186
   knives, 163–167
   paraffin, 162–163
   rotary, 163
   Schantz, 162

Microtome, (*See also* Freehand sections; Frozen sections; Paraffin sections)
Millimeter equivalents, 13
Mite, wholemount of, 197–199
Mitochondria, method for, 280
Mitosis, method for, 279, 286
   in rat testis, 253
Mohr and Wehrle's mountant, 126
   for blood smear, 221
Monocot, leaf, 282
   stem, 280
Monocystis, smear of, 216–218
Mordant, explanation of, 98
   staining, 99–101
Motor cells, method for, 278
Motor nerve ends, method for, 279
Mounting, Amphioxus section, 258–260
   Aristolochia stem, 248
   bacteria, 222–229
     in sections, 261–263
   blood, 219–221
   chick embryos, 206–211
     section, 264–267
   freehand sections, 149
   frozen sections, 189–190
   intestine, 237–244
   liver fluke, 212–215
   mite, 197–199
   monocystis, 216–218
   paraffin blocks, 168–170
     sections, 174–183
   Pectinatella, 200–205
   root, 233–236
   salivary gland, 230–232
   testis, 250–254
   tongue section, 255–257
Mouse, method for cerebrum, 271
Murphy (*see* McNamara, Murphy, and Gore's solution)
Muscle, methods for, 278
Mycorrhiza, method for, 284
Myofibrils, method for, 280

N.A. (numerical aperture), control of, 16
   effect of, 16–20
   explanation of, 8–14

Narcotization, 95–96
  bryozoa, 201
  invertebrates, 131–133
  liver fluke, 212
Navashin's fixative, 89
Necturus, method for liver, 274
Neelsen technique, 227–229
Negative, photographic, characteristics, 56–58
  processing, 61–64
Nervous system, methods for, 271–272
  tissues, methods for, 278–279
Neuroglia, method for, 278
Neutral density filter, 24
"Neutral" mountants, 126
Nissl bodies, method for, 280
Nitric acid, for decalcifying, 93–94
  in fixative, 89, 91
Nonsilica glass, 8
Nuclear stains, explanation, 98
Nuclei, methods for, 279–280
  stains for, 98–109
Nucleic acid, 98
Numerical aperture (*see* N.A.)

Objective, achromatic, 44
  apochromatic, 44
  corrections of, 14
  explanation of, 4
  for freshman microscope, 32
  magnification, 12–13
  for medical microscope, 35
  mm equivalents, 13
  for research microscope, 44
  resolution, 8–14
  types of, 14–15
  water immersion, 253
  working distance, 14
Ocular, explanation of, 4
  eye relief, 4
  pinhole, 49
  types of, 15
Oil blue N, 116
Olfactory pit, method for, 272
Oligochaetae, wholemounts, 133
Oocytes, method for, 275

Optic lobe, method for, 271
Orange G, 115
  for Aristolochia, 248
  staining with, 111
Orange II, 110
Orcein, 105–106
  for chromosomes, 230–232
Organelles, methods for, 280
Organs, animal, methods for, 271–276
Oribatid mites, 198
Ortho-Illuminator, 21, 43, 47
Osmic acid, 86
  applied to smears, 143
Ovary, invertebrate, method for, 275
  plant, method for, 282
  vertebrate, method for, 275
Ovens, embedding, 154–155
Oviduct, method for, 276
Oxalic acid, in carmine staining, 215
  as ingredient in stain, 111

Panchromatic material, 58
Pancreas, method for, 274
Papers, photographic, 59–60
Paraffin ribbons (*see* Paraffin sections)
Paraffin sections, 149–185
  action of knife in cutting, 165
  attaching to slide, 172–177
  clearing, animal material for, 151
    plant material for, 151–152
  combining single into ribbons, 265
  cutting, 165, 170–185
  defects appearing, after mounting, 183–184
    while cutting, 178–180
  dehydration of material for, 151–152
  embedding, media for, 152
    technique of, 153–162
  flattening, 175–177
  knives for, 163–167
  making blocks for, 159–162
  microtomes for, 162–163
  mounting blocks for cutting, 168–170
  ribbons of, 172–177
  outline of technique for, 79, 149
  preparing embedding boxes for, 156–158

Paraffin sections, rolling to slide, 177, 256
  selection of fixative for, 150
  sharpening knives for, 164–168
  staining, 180–185, 241–244
Parathyroid, method for, 274
Parenchyma, method for, 284
Patay's stain, 112
Pectinatella, wholemount of, 200–205
Penis, method for, 276
Permeability, effect on staining, 98
Petroleum jelly for temporary wholemounts, 129
Petrunkewitsch's fixative, 91
  for rat tongue, 255
Phase, relation to amplitude, 22, 27
Phase-contrast, explanation, 27–28
  illumination, setting up, 53
  microscope, setting up, 51–54
  substage condenser, 52
Phase plate, 28
  centering, 53
Phase shift, 26
Phellem, method for, 283
Phellogen, method for, 283
Phenol, for clearing, 121
  in fixative, 91
  for softening chitin, 94
  in stain, 108
  for swelling dried plant specimens, 233
Phenyl salicylate as ingredient in mountant, 126
Phloem fibers, method for, 283, 285
Phloroglucinol in decalcifier, 93
Phloxine, 114, 262
Phosphomolybdic acid for differentiating, 112
Phosphotungstic acid differentiating solution, 111, 112
Photographic dark room, 60
Photographic exposure, 72
Photographic filters, 71–72
Photographic negatives (*see* Negatives)
Photographic prints, processing, 64–66
  selection of paper for, 59–60
Photography, principles of, 55–56
Photomicrograph, taking, 69–73
Photomicrographic cameras, 66–69

Photomicrography, contrast in, 58–59, 72
Picric acid, for differentiating stains, 100
  as fixative, 90–91
  removal from tissues, 92–93
  as stain, 110–111, 256
Pigeon, method for, pancreas, 274
  tendon, 277
Pigment cells, method for, 278
Pineal, method for, 272
Pipette, definition of, 82
Pistil, method for, 282
Pith, method for, 282, 283
Pits, method for, 285
Pituitary, method for, 274
Placenta, method for, 276
Placoid scale, method for, 271
Plant chromosomes, 89
Plant organs, methods for, 280–282
Plant ovary, method for, 282
Plant tissues, dehydrating and clearing, 151
  freehand sections of, 147
  methods for, 282–284
  wholemounts, 134
Plasmodesmata, method for, 286
Platyhelminthes, wholemounts, 132
Polychaetae, wholemounts, 89, 132
Polyethylene staining jars, 84
Polyvinyl alcohol as mounting medium, 125
Ponceau 2R, 110, 112
Potassium alum in carmine stain, 105
Potassium dichromate in fixatives, 87, 91
Potassium hydroxide for softening chitin, 198
Potassium permanganate for bleaching carmine, 215
Potato, method for stem, 281
Preservation of cellular detail, 86–87
Prints, photographic, contrast of, 59
Prism, 5
Pronephros, method for, 276
Prostate, method for, 275
Protein, fixation of, 86–87
Prothallium, staining with carmine, 105
Protozoa, staining, 101
  wholemounts, 132
  temporary, 129
Pseudonavicellae, 217

Purkinje cells, method for, 278
Pyramidal cells, method for, 278

Quillwort stem, method for, 281

Rabbit, method for, alimentary canal, 273
　　connective tissues, 277–278
　　ear, 277
　　glandular system, 274–275
　　heart, 275
　　liver, 261–263
　　lung, 273
　　nervous tissues, 278–279
　　ovary, 275
　　reproductive structures, 275–276
　　spinal cord, 272
　　tongue, 271
Ramsden's disc, 4
Ranunculus, method for root, 281, 282
Raphids, methods for, 285
Rat, method for, alimentary canal, 272–273
　　kidney, 276
　　salivary gland, 274
　　testis section, 250–254
　　tongue section, 225–257
Rectum, method for, 273
Reflex image projection, 67, 79
Refraction, 5
Regaud's hematoxylin, 100
Renal structures, methods for, 275–276
Reproductive structures, methods for, 275–276
Reptile scale, method for, 271
Resin for differentiating stain, 114, 262
Resin ducts, method for, 284
Resinous mounts, 126–127
Resolution, 8–12
　　control of, 16–20
　　of photographic negatives, 57
　　relation to angular aperture, 11
　　table of, 13
Respiratory system, methods for, 273
Resting nuclei, method for, 279
Retina, method for, 279
Ribbons, paraffin (see Paraffin sections)

Richards, O. W., 177, 190
Ringer's solution, 144, 230
Roll film processing, 63–64
Root, methods for, 281
　　section of, 233–236
Rotifera, 85, 95
Rubber paraffin, 152

Safranin, 107, 114
　　for bacteria, 226
　　for plant tissues, 248
Salicylic acid in balsam, 112
Salivary gland, method for, 274
Salol as ingredient in mountant, 126
Sass, J. E., 287
Scales, methods for, 271
Schantz microtome, 162
Sclereids, method for, 285
Sclerenchyma, method for, 284
Sea anemone, narcotizing, 95
Secondary root, method for, 281
Secretory cells, plant methods for, 284
Section, of Amphioxus, 258–260
　　of intestine, 237–244
　　of liver, 261–263
　　of plant stem, 245–249
　　of rat tongue, 255–257
　　of root, 233–236
　　of 72-hour chick, 264–267
Section lifter, 82
Sections, diagram of technique for, 79
　　standard planes for, 145
　　techniques (see Freehand sections; Frozen sections; Paraffin sections)
Sedum, method for leaf, 281, 282
Seeds, fixative for, 90
Sensory tissues, methods for, 278–279
Serratia marscens, 70
Setting up, freshman microscope, 32–34
　　immersion lens, 40, 50
　　medical microscope, 36–42
　　research microscope, 47–51
Sharpening microtome knives, 165–168
Shrinkage in fixation, 86
Sieve cells, method for, 285
Sieve-tube elements, method for, 285

Silge and Kuhne, 43
Sine wave representing light, 22
Skeletal structures, methods for, 273–274
Skeleton, arthropod, method for, 274
Skeletonizing plant tissues, 234
Skeletons, staining, 116
Skin, method for, 271
Slides, cleaning, 139–141, 191–192
    jars for staining, 83
    labeling, 192–193
    specifications of, 80
    storage, 193–194
    thickness, effect on substage condenser, 51
Smears, of bacteria, 222–224
    of blood, 219–221
    diagram of technique for, 78
    fixing, 143
    of monocystis, 216–218
    staining, 144
    technique of preparing, 140–144
    of tubercle bacilli, 227–229
Smith's stain, 111
Smooth muscle, method for, 278
Sodium acetate as ingredient of stain, 107
Sodium bicarbonate for bluing, 102, 103
Sodium hypochlorite, for hydrolyzing sputum, 228
    for skeletonizing plant tissues, 234
Sodium sulfate, in fixative, 87, 88
    in photographic hardener, 62
Sodium sulfite for macerating wood, 95
Sodium thiosulfate in photography, 62
Softening chitin, 198
Solubility, effect on staining, 99
Solvents, "universal," 119
Sparrow, method for, cerebellum, 271
    eye, 272
Spectral sensitivity of photographic negatives, 57
Speed, of papers, 59
    of photographic films, 57
Spencer microtomes, 163, 180
Spermatogenesis in rat testis, 250–254
Sphagnum moss, collecting small animals from, 197–199
Spherical aberration, 6

Spinal cord, method for, 272
Spirit blue, 111
Spleen, method for, 275
Split-beam image, projection of, 66, 69
Sponge spicules, method for, 274
Spore cases of monocystis, 217
Sputum, bacteria from, 227–229
Squamous epithelium, method for, 277
Squashes, diagram of technique for, 78
    freezing to coverslip, 232
    of salivary gland, 230–232
    stain for, 106
    technique of, 144
Stage, for microscope, 43
    mechanical, centering, 51
Staining, with acid fuchsin, 108, 110
    with alizarin red S, 116
    with aniline blue, 111
    with azure II, 114, 262
    bacteria, with crystal violet, 223
        with safranin, 226
        in sections, 114
    blood, 113–114
    bone, 113, 116
    with carmine, 104–105, 200–205
    cartilage, 112, 113, 116
    celestin blue B, 106–107, 242, 256
    chick embryo, 210–211
    chromosomes, 105–106, 115
    connective tissue, 111–113
    with crystal violet, 108–109
    direct, 98–99
    effect of charge on, 98
    with eosin, 109, 243
    with fast green, 109, 115, 245
    fats, 116
    freehand sections, 149
    with hematoxylin, 99–104, 266
    indirect, 98–99
    keratin, 111
    with light green, 112
    liver fluke, 214–215
    with magenta, 108, 228
    with methyl blue, 259
    with methyl violet, 115, 248
    with methylene blue, 113, 116
    mordant, 99–101

Staining, muscular tissues, 110–112
  nuclei, 98–109
  with oil blue N, 116
  with orange G, 111
  with orange II, 110, 259
  with orcein, 105, 106, 230–232
  paraffin sections, 180–185, 241–244
  parasitic fungi, 115
  with phloxine, 114, 262
  with picric acid, 110–111, 256
  plant sections, 114
  plasma, 109–111
  Ponceau 2R, 110
  principles of, 97
  protozoa, 101
  purpose of, 97
  with safranin, 107–108, 114–115, 248
  skeletons, 116
  small invertebrates, 113
  smears, 144
  solutions (see name, e.g., Wright's stain, or ingredient, e.g., Eosin)
  with spirit blue, 111
  squashes, 106
  with sudan IV, 116
  with toluidine blue, 116
  vertebrate embryos, 131
  xylem, 103, 155
  yolky material, 111
Staining jars, 83
Stamen, method for, 282
Starch grains, method for, 286
Steam for fixing amebas, 132
Steedman, H. F., 190
Steles, methods for, 283
Stem, methods for, 280–281
Stender dish, 82
Stentor, method for myofibrils, 380
Stomach, method for, 272
Stone cells, methods for, 285
Stop, annular, 28
  central, 29
Storage, of paraffin blocks, 161
  of slides, 193–194
Stratified epithelium, method for, 277
Striped muscle, method for, 278
Stropping microtome knives, 167

Substage condensers, 16–20, 36
  centering, 48
  effect on resolution, 10–11
Substage turret, 52
Succulent leaf, method for, 282
Succulent stem, method for, 281
Sudan III, 99
Sudan IV, 116
Supplementary lens with substage, 41
Supporting media, for freehand sections, 147
  for frozen sections, 186
Surface of photographic papers, 60
Syracuse watch glasses, 81–82
Syrup in embedding medium, 186

Tadpole, method for, chromosomes, 279
  kidney structures, 276
  skin, 276
Teleost scale, method for, 271
Tendon, method for, 277
Terpineol for clearing, 120, 135
Tert-butyl alcohol, for dehydrating plant materials, 151, 248
  for dehydration, 150
  in Johansen's stain, 115
Testis, method for, 276
  section of, 250–254
Thymus, method for, 724
Thyroid, method for, 274
Thysanura, 198
Tissues, animal, method for, 277–279
  plant, methods for, 282–284
Toluene for clearing tissues, 121–122
Toluidine blue, 116
Tongue section, 255–257
Tooth, method for, 271
Tracheid, method for, 285
Trichloroacetic acid, for decalcification, 94
  in fixative, 91
Trichomes, method for, 284
Trinitrophenol (see Picric acid)
Trinocular microscope, 67–69
Triple stains, 111–114
Tuber, method for, 281
Turret substage, 52
Typha, method for stem, 281

Undeviated wave, 22, 24–29
"Universal" solvents, 119
Uranyl acetate in fixative, 91
Urea in fixative, 91
Ureter, method for, 276
Useful magnification, 13
Uterus, method for, 276

Vagina, method for, 276
van Gieson's stain, 110
  for rat tongue, 256
van Wijhe's stain, 116
Vas deferens, method for, 276
Vascular cambium, method for, 284
Vascular cylinder, methods for, 283
Vessel plants, method for, 285
Vials, 82
Vicia seedlings, method for root, 281
Vortex, wholemount, 132

Washing photographic negatives, 62
Watch glasses, 81–82
Wave, deviated, 22, 24–29
  undeviated, 22, 24–29
Wavelength of light, 22
  effect, on dispersion, 5
    on focus of lenses, 6
    on photographic plate, 57
    on resolution, 9
Wess (*see* Gray and Wess's medium)
White fibrous tissue, method for, 277
Wholemounts, diagram of technique for, 78

Wholemounts, in gum media, 129–130
  of invertebrates, 132–133
  of liver fluke, 212–215
  of mite, 197–199
  of Pectinatella, 200–205
  in resinous media, 131–139
  temporary, 128–129
  of thick objects, 130
  of 33-hour chick, 206–211
Willow, method for stem, 280
Wood, method for, 285
  softening, 94
Working distance, 14
Wright's stain, 113
  for blood smear, 219–220

Xerophyte stem, method for, 281
Xerophytic leaf, method for, 282
Xylem, 115
  methods for, 283
Xylene, for clearing, 121, 151
  as solvent for balsam, 127

Yellow elastic tissue, method for, 277

Zenker's fixative, 87
  for liver, 261
  for oligochaetes, 133
Ziehl-Neelsen technique, 227–229
Ziehl's carbolmagenta, 108
Zimmerman's lacquer, 189